RF and Time-domain Techniques for Evaluating Novel Semiconductor Transistors

Keith A. Jenkins

RF and Time-domain Techniques for Evaluating Novel Semiconductor Transistors

 Springer

Keith A. Jenkins
Technical Consultant
New York, NY, USA

ISBN 978-3-030-77777-7   ISBN 978-3-030-77775-3   (eBook)
https://doi.org/10.1007/978-3-030-77775-3

This Springer imprint is published by the registered company Springer Nature Switzerland AG
The registered company address is: Gewerbestrasse 11, 6330 Cham, Switzerland

Preface

The unrelenting demand for faster and more powerful computing and communication in smaller devices and computers has led to a many decades-long continuous improvement in the transistors, mostly field-effect transistors (FETs) which are the building blocks of computers. Much of the progress has been achieved by reducing the size of the FETs, which has been accomplished by the methods known as "scaling." However, the prospect of continued scaling has diminished in the last few years, and since the demand for performance is continuing to increase, there is considerable interest in finding other ways to improve transistor speed, and, if possible, to do so without requiring a drastic increase in power consumption.

This has led to many interesting proposed replacements of conventional transistors, by significantly modifying their structures, or by replacing them entirely with devices made of new materials. With all of these proposed replacements comes the important task of evaluating their performance: in additional to measuring DC currents and voltages, it is incumbent on their inventors to demonstrate that they switch or amplify quickly, are reliable, don't degrade with time, have predictable performance have measurable at-speed characteristics, can be represented in an equivalent circuit model which can be used for circuit design.

This book is concerned with presenting ways to examine these performance questions, to go beyond DC measurements to measure dynamic effects. Drawing from common knowledge, published literature, and from the author's own investigations, it describes a number of measurement techniques which can be applied to novel devices. It covers frequency-domain and time-domain operation, and small-signal and large-signal operation. It describes some well-established measurement methods, as well as a number of novel techniques which have been developed as a result of innovation in transistor structure. The book could be described as presenting ABDC techniques: Anything But DC.

Throughout the book, the emphasis is on the practical aspects of measurements, as actual practice of the skill in the lab, even for well-established methods, is seldom as easy as theory suggests. Math is kept to a minimum, and only results are stated, without derivations. Along with the descriptions of measurement methods, illustrative examples are presented, when possible, of the use of the methods applied to a variety of physically realized novel transistors. The emphasis, though, is on the measurement techniques and their relevance, not on the physics of the illustrative transistors.

The book is intended for readers, such as students and academic and industrial researchers, who need to understand the dynamic behavior of transistors and the potential impact of that behavior on the performance of circuits. The reader is assumed to understand elementary linear electrical circuit theory, is familiar with conventional silicon transistors (mostly FETs) and their DC characteristics, has experience with measurements using DC equipment, and is acquainted with AC signal concepts.

New York, NY   Keith A. Jenkins

Acknowledgments

The author would like to thank the following people for contributing to this work through their collaboration, technical discussions, management support, encouragement, and friendship:

Paul Agnello, Joerg Appenzeller, Phaedon Avouris, Karthik Balakrishan, Aditya Bansal, Steven Bedell, Manjul Bhushan, J.N. Burghartz, Jin Cai, Eduard Cartier, C.-T. Chuang, T.C. Chen, George Chiu, H.-Y. Chiu, John Cressler, Bijan Davari, Christos Dimitrakopoulos, Michael Engel, Damon Farmer, Robert Franch, David Frank, Jacques Gautier, Shu-Jen Han, Dave Heidel, Mark Ketchen, Marwan Khater, Jae-Joon Kim, Steven Koester, Siyu Koswatta, G.-P. Li, Yu-Ming Lin, Barry Linder, Shih-Hsien Lo, Pong-Fei Lu, Inanc Meric, Hwall Min, Vijay Narayanan, Tak Ning, Modest Oprysko, Mario Pelella, Jean-Luc Pelloie, Mario Pelella, Stas Polonsky, Rahul Rao, Ken Rim, Devendra Sadana, Ghavam Shahidi, Eugene Shapiro, Melanie Sherony, Dinkar Singh, Paul Solomon, Peilin Song, Kevin Stawiasz, Franco Stellari, Matthias Steiner, Johannes Stork, Lisa Su, J. Y.-C. Sun, C-Y Sung, Denny Tang, Jianshi Tang, Jamie Teherani, Bharath Takulapalli, Alberto Valdes-Garcia, Rick Wachnik, Clement Wann, Alan Weger, H.S. Philip Wong, Yanqing Wu, and J. B. Yau.

Contents

About the Author

Keith A. Jenkins was a Research Staff Member at the IBM Thomas J. Watson Research Center from 1983 to 2018. In this position, he had the privilege of working with device development, technology evaluation, and circuit design, leading to research in a large variety of device and circuit subjects. These include high-frequency measurement techniques, electron beam circuit testing, radiation-device interactions, low temperature electronics, SOI technology, substrate crosstalk in circuits, frequency response of nanoscale devices, and studying the impact of self-heating in advanced CMOS technologies. He also worked to design circuits for analog on-chip self-measurement, including jitter and phase error measurement, and on-product reliability monitoring. He designed several compact and efficient structures to measure device performance, uniformity, and device reliability, in order to replace the discrete transistor structures usually used for these studies.In pursuing these research goals, he developed many new measurement techniques, some of which are covered in this book. He received several technical awards from IBM, and several best paper awards from conferences and journals. He has published more than two hundred research papers and has been granted more than thirty U.S. patents.

He received a PhD in physics from Columbia University for experimental work in high energy physics, and before joining IBM, he worked in high energy physics at The Rockefeller University. He was an adjunct professor of physics at Hunter College and Manhattan College.

Dr. Jenkins is now engaged as a consultant on the subject of semiconductor device and circuit measurements.

Chapter 1
The Novel Transistor Characterization Challenge

1.1 Introduction: Why This Book?

Transistors are the building blocks of all modern electronic devices, which span all parts of modern life from cell phones to PCs to massive transaction centers like banks and airline reservation systems. The dominant commercial transistor is the silicon-based MOSFET (metal-oxide-silicon field-effect transistor), which is fabricated as *n*-type and *p*-type (nFET and pFET). The dominant digital circuit technology is that of CMOS (complementary metal-oxide-silicon) which uses *n*- and *p*-MOSFETs. As wildly successful as it has been in modern electronics, however, this silicon technology has reached a point where conventional electronics, based on the integration of millions or billions of CMOS FETs may not have much prospect of improvement, yet there is continual demand for greater speed and higher density, amid concern about power consumption. As a result, there is a great interest in developing new transistors to replace or supplement conventional silicon CMOS. Some new transistors may be developed through changes from modifications to standard CMOS, such as new gate materials, new gate dielectrics, insulating substrates, non-planar structures, and alternate materials, and others may come from the use of totally new materials, such as two-dimensional crystals, for the conducting channels. Either way, the new transistors will use different materials and structures, and as such, their electrical properties may be quite different from standard CMOS. The term semiconductor, in the title of the book, is used with admitted sloppiness. Some new materials from which novel transistors are made, are semimetals, not semiconductors, but all active transistors make use of the fact that their conduction can be modulated by a controlling input signal.

This book asks the questions: how do we know if these new transistors are any good, and how well do they work? Field-effect transistors, with which this book is almost exclusively concerned, are actually transconductance (not transresistance) devices, as their current output is a function of voltage input. Simple static DC measurements, *e.g.*, *I–V* curves of output conductance and voltage transfer, and low-frequency capacitance measurements provide almost all of the information needed for conventional MOSFET characterization, and for developing complex equivalent-circuit models, such as SPICE, which are used for circuit simulation.

© Springer Nature Switzerland AG 2022
K. A. Jenkins, *RF and Time-domain Techniques for Evaluating Novel Semiconductor Transistors*, https://doi.org/10.1007/978-3-030-77775-3_1

To answer these seemingly simple questions above, several conventional characteristics may be required. For example, depending on the intended use of the transistors, such as analog or digital circuits, it might be desirable to measure

- small-signal frequency response: cutoff frequency, maximum frequency of oscillation
- small-signal linearity
- digital switching speed

These parameters can be measured using well-known techniques, which are presented in the book, but the techniques assume that the transistor behavior is unchanged during its operation. However, since it is possible that novel transistors may not be as stable and reliable as those of conventional silicon technologies, dynamic measurement techniques may be required for their characterization. For example, some effects which novel transistors may exhibit are

- self-heating
- history-dependent output
- very short-term degradation
- DC-invisible structural defects
- different response to DC and transient inputs

None of these effects would be apparent in traditional DC characterization measurements and all would give different assessments of the quality and performance of the transistors. The book presents a variety of techniques using high-frequency ("RF") and time-domain measurements to better understand the electrical performance of devices which may be subject to such effects.

Consider an example. A new material is discovered which has some attractive electrical properties, of which the most important is a very high electron (or hole) mobility when measured in its pure form. Experiments show that the concentration of carriers in the material can be modulated by the application of a surface voltage. Thus, the material seems like a good candidate with which to build an FET, one which may switch or amplify faster than silicon MOSFETs.

A prototype device which has a structure similar to a planar silicon MOSFET using this material as the channel is built, and when tested with static I–V methods, shows transfer characteristics (I_d – V_g curves) and output characteristics (I_d – V_d curves) which have some similarity to a MOSFET. While I–V curves can give an indication of channel mobility, it is of natural interest to see if the frequency response of the transistor is as high as was expected from the known high mobility of the starting material. This obviously requires some sort of gain vs frequency measurement, which may or may not be conventional.

Even if simple gain measurement suggests good performance at high frequency, the transistor is being considered as a replacement for standard MOSFETs; it is important to determine if it truly behaves like a transconductance device in which the frequency is governed by a gate capacitance and DC transconductance. This requires comparison of high-frequency response to a standard small-signal (to be defined in Chap. 3) "hybrid-pi" model.

Suppose this new transistor also has two peculiarities. (1) The transconductance determined by the static curves is much smaller than it should be based on the measured mobility of the pristine material. (2) The $I-V$ curves are not stable, that is, two curves taken in succession give different results, and there is hysteresis when the voltage sweep is reversed. The frequency response, measured at constant gate and drain voltages, confirms that the mobility is lower than expected. This observation by itself may suggest that the material has lost some mobility during its fabrication, and the hysteresis suggests that something changes in the channel or dielectric when the transistor operates and current flows through it. Taken together, these might suggest that in addition to loss during fabrication, when voltage biases are applied, there is some rapid degradation of the channel material, or charging at the interface between the channel and the gate dielectric, and the time scale for this degradation is much smaller than the time it takes to measure the static curve. If this can be understood, then it might lead to the use of a different dielectric, or a different transistor structure to build a device with higher DC transconductance. To explore this possibility requires some technique, or techniques, to determine if the short-time response of the transistor differs from the DC response.

In the following chapters, these sorts of questions are addressed. Using examples from conventional and exploratory transistors, a variety of high-frequency and short-time domain evaluation techniques are presented and explained. It is not expected that all, or even most, will be required for every new kind of device, but having a collection of possible techniques available should help in understanding the dynamic electrical performance of new kinds of transistors.

1.2 General Measurement Conditions and Terminology

The evaluation of a transistor's performance is a matter of applying a stimulus signal to it, and measuring its response, that is, its output. That description is, of course, as true for conventional DC characterization as it is for the techniques which are presented in this book. Note that in this book, the term "signal" refers to any form of measurable variation of an input or output electrical quantity. Many RF engineers regard the word as meaning specifically a voltage sine wave, but this book uses the term much more broadly. Because the operating frequency of today's transistors and circuits is very high, the measurement techniques in this book may have to use correspondingly high-frequency signals, both for stimulus and response.

However, to go a little deeper into what is to be learned it is useful to consider what are the measurement conditions and the nature of the signals, and to clearly distinguish among them, to provide clear terminology for the discussion. Some of these measurement conditions are presented below.

1.2.1 Small-Signal vs Large-Signal

Unlike resistors and capacitors, most semiconductor devices are nonlinear, that is, their outputs are not linear functions of their inputs. This leads to the distinction between large- and small-signal. This distinction is important in DC measurements as well as in the techniques presented in this book. Large-signal usually, but not necessarily, refers to signals which switch between the transistor's limiting voltages, commonly known as rail-to-rail operation. For example, if a CMOS FET is operated with a 0.8 V power supply value and a 0.0 V ground, a large-signal on the gate would be a change from 0 V to 0.8 V. If applied to a CMOS inverter, the change in gate voltage would cause the output to switch from 0.8 V to 0 V, also a large-signal swing. During the switching of a single FET, the current through the FET would vary with the gate voltage in a very nonlinear fashion: when the gate voltage is below threshold voltage, the drain current is very small, but with an exponential dependence on voltage, and above the threshold, the drain current might change in proportion to the square of the difference between the gate voltage and the threshold voltage (often called the overdrive voltage).

Small-signal operation is the situation when the swing of the input and output are reduced to a level equal to or lower than the amplitude at which the transistor responds linearly to the input. From a DC perspective, small-signal refers to differential values and can be obtained from swept measurements by differentiation. The difference between differential resistance and total resistance of a diode is an example of this distinction. The static total value is obtained from the total voltage and current, and the differential resistance is obtained from the derivative of voltage with respect to current, i.e., the ratio of small-signals. In the treatment of time-dependent operation, DC voltages set the operating point, often called the quiescent point, of the transistor which can be represented in a model as a network of linear components to which small-signals are applied, superimposed on the DC voltage. This criterion for linear representation can be estimated by taking a Taylor series expansion of the I–V formula, and determining what is required to ignore all the terms of order higher than linear.

For example, in the saturation region of a classical long-channel FET, the drain current can be simplified as depending on the gate voltage by

$$I_d = A\left(V_g - V_t\right)^2 \tag{1.1}$$

where V_g is the gate voltage, V_t is the threshold voltage, and A is a constant which represents multiple device-specific constants as one number. Then if ΔV_g is the change of gate voltage, using a Taylor series to estimate the change of drain current resulting from this gate voltage change,

$$I_d\left(V_g + \Delta V_g\right) = I_d\left(V_g\right) + 2A\left(V_g - V_t\right)\Delta V_g + A\Delta V_g^{\ 2} \tag{1.2}$$

and the series terminates. The small-signal approximation requires that the nonlinear terms, which in this case is purely quadratic, be much smaller than the linear term, or

$$\Delta V_g \ll 2\left(V_g - V_t\right) \tag{1.3}$$

If this is the case, then the gate voltage swing, ΔV_g, is linear. This is easily satisfied, even for fairly large voltage differences, except when the gate voltage is near-threshold voltage. Below threshold voltage, the *I–V* curve becomes exponential:

$$I_d = Be^{qV_g/kT} \tag{1.4}$$

where q is the charge of the electron, k is the Boltzmann constant, T is the temperature, and kT/q is the thermal voltage. A Taylor series expansion of Eq. (1.4) leads to

$$\Delta I_d \cong I_d\left(V_g\right)\left(\frac{q\Delta V_g}{kT} + \frac{1}{2!}\left(\frac{q\Delta V_g}{kT}\right)^2 + \dots\right) \tag{1.5}$$

The quadratic and higher order terms can be neglected when

$$\Delta V_g \ll \frac{2kT}{q} \tag{1.6}$$

Thus linear, small-signal, operation is obtained in subthreshold operation when the gate voltage increment is much less than the thermal voltage, which is a much more stringent requirement than in the saturation region. The definition of small-signal depends on the quiescent point. In practice, there is not much dynamic information to be learned from operation of the transistor in the subthreshold region, and it is probably adequate to use a gate signal which is comparable to, or slightly less than, the thermal voltage (26 mV, or about −23 dBm (rms) in a 50 Ω resistor, at room temperature). In practice, it is always possible to test if the transistor is operating in the small-signal regime by changing the input signal amplitude and seeing if the output changes in proportion to the input. For novel devices for which a mathematical description is not yet established, experiment might be the only way to determine the small-signal level. The drawback of using very small input signals is that output signals are also likely to be small, and then measurement noise is relatively greater, so experimentally finding the largest acceptable input which is still small-signal is a useful procedure.

In this book, large-signal voltages and currents are indicated by upper case letters; small-signals and differential quantities, lower case. Thus, the total voltage applied to an FET gate by a small-signal sine wave is written as $V_g + v_g \, sin(\omega t)$, where V_g is the quiescent bias point of the gate and v_g is the small-signal amplitude of the sine wave.

1.2.2 Frequency-Domain vs Time-Domain

In DC measurements, the input and output currents and voltages are essentially static. Even though in the measurement of DC I–V curves, the biases are stepped over their ranges, the measurements at each step are made in a static condition (and the engineer is given control of delays in order to guarantee it.)

But in the techniques described in this book, inputs are deliberately time varying, in order to observe dynamic effects of the transistors. The variation of inputs can be in the frequency-domain, in which a sine wave, or several sine waves, is applied, or in the time-domain, where a voltage is varied according to a timed pattern. Similarly, outputs can be measured in either domain. Although it is commonplace to regard time and frequency as variables which can be transformed back and forth by a Fourier transform, in practice they are quite different. For example, a single very short pulse can be applied to a device, but a single very short sine wave, if it is shorter than the period, is not even meaningful concept. The use of the word "RF" in the title of the book refers to the general role of high-frequency signals,[1] not specifically to the frequency band which it designates for communications circuits, but generally encompassing signals of hundreds of MHz to tens of GHz.

Measurements in both domains are useful and practical and should be considered according to the problem being addressed. But in general, frequency-domain measurements tend to be used with small-signal stimuli to understand the network nature of the transistors. This is the arena of analog and RF circuits. Time-domain measurements tend to be used in large-signal situations and are useful for understanding the response of transistors to full, rail-to-rail voltage inputs. This is the arena of digital circuits.

1.2.3 Continuous-Wave vs Transient Signals

Finally, consideration must be given to the persistence of the stimulus or stimuli. The term continuous-wave (CW) is used to describe the situation where a time-varying signal is applied continuously. This would be the case of the sine wave (in the frequency-domain) and square wave (in the time-domain) shown in Fig. 1.1a, b. The signals are regarded as having always been present and continuing forever. Such a situation usually leads to the transistor being in a steady-state condition.

---

[1] In communications language, "high frequency" can also designate a particular frequency band, one which would be considered rather low frequency in today's world.

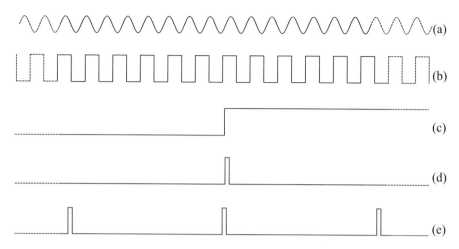

Fig. 1.1 Illustrations of various input signal types used for testing transistors. (**a**) CW sine wave, (**b**) CW square wave, (**c**) transient step, (**d**) transient pulse, (**e**) repetitive transient pulse

(Although long time operation can lead to small effects such as hot-carrier induced degradation of threshold voltage and mobility changing the steady-state condition slightly.)

But in many circumstances, response of the transistor immediately after application of the input signal is more important than the steady-state operation. The extreme example of this is the CMOS logic gate. Most of the time, the gates are in a static DC state, at fixed high or low voltage, and only conduct current for a very short time, a few picoseconds, when their logic state changes. In a typical microprocessor, for example, most gates switch less than 1% of the time. The current available during that brief time is what determines delay of the circuit, not the current which would be present in CW operation. This brings out the possible importance of transient operation, illustrated in Fig. 1.1(c), where the signal is switched once from one value to another, that is, it has a transient *step*. Complete transient response observation would measure the current development immediately after the voltage step and continue for as long a time as desired.

Figure 1.1(d) illustrates another possibility, in which a single pulse is applied for a very short time. In the language of voltage steps, this is actually a positive step followed by a negative step, separated by a short time, but it just as well can be regarded as a single transient *pulse*.

As will be discussed in Chap. 2, most of the measurement equipment used to measure frequency- and time-domain outputs of transistors requires repetitive signals which persist for a reasonably long time, so measurement of transient events is fairly challenging. However, the situation of Fig. 1.1(e) can sometimes be used to determine transient operation. Figure 1.1(e) describes *repetitive* transient pulses though not to scale. Unlike the CW operation of Fig. 1.1-(b), the short pulses are separated by a long time, and if, and only if, the transistor returns to its initial state in the time between pulses, a repetitive pulse train can be used to measure response to a transient pulse. It is also possible to apply pulsed frequency-domain signals, in which relatively short burst of sinusoids are applied, but this situation is not considered in this book.

1.3   Overview of the Content of the Book

The book presents a collection of RF and time-domain techniques used to answer some of the performance and characterization questions raised above. It uses examples of experimental results from published works using a variety of novel devices. The progression of the techniques presented is from predominantly small-signal frequency (RF) characterization to large-signal time-domain characterization.

While describing a variety of measurement needs and solutions, new measurement problems are always cropping up, and there is no claim that the book is exhaustive. Rather, it is hoped that understanding the techniques which have been developed may lead to new methods, or improvement on existing ones, in the hands of inquisitive investigators.

Chapter 2. High-Frequency Test Equipment, Connections, and Contact with Transistors

This chapter describes some of the equipment and apparatus needed for the measurement techniques which are explained in the rest of the book, with emphasis on spectrum analyzers, vector network analyzers, and oscilloscopes. Attention is paid to resolution and dynamic range of the instruments. The importance of transmission line effects of coaxial interconnection between the equipment and the transistors is discussed in some detail. This chapter may be skipped initially, and used later as a reference, by readers with some familiarity with test equipment and transmission line properties, but is presented first for those who are unfamiliar with the subjects.

Chapter 3. Frequency Response and Gain

This chapter is concerned with how to assess the frequency response of transistor. The design of transistor structures suitable for high-frequency measurements is discussed. Several methods are discussed by which the gain as a function of frequency can be measured for transistors with very low output current for which traditional techniques may not apply. Then the traditional method of S-parameter measurement is explained, followed by conversion to figures-of-merit. The hybrid-pi small-signal model is used for comparison with measured parameters. Measurement structures, network analyzer calibration and de-embedding of transistors from probe pads are explained in detail.

Chapter 4. Case Studies in the Evaluation of Novel Transistors

This chapter treats two separate unconventional problems which can be addressed using frequency-domain methods. The first is the measurement of voltage gain as a function of frequency for the situation where the novel transistor doesn't provide enough current to deliver power gain, but is still of interest as a high-frequency voltage amplifying transistor. The second is concerned with the diagnosis of high resistance in the gate of an FET. By using a vector network analyzer, it is demonstrated that the gate may have a high vertical resistance which cannot be seen in DC measurements and which may degrade circuit performance. The frequency method identifies and measures this vertical resistance.

Chapter 5. Measurement of the AC Linearity of Transistors

This chapter takes up the assessment of the nonlinearity of transistors. Usually applied to analog circuits, transistor linearity measurement can be useful in order to establish a starting point from which to build linear circuits, and to get insight into the transistor's transport mechanism. Two types of measurements of nonlinearity are described, and the connection between nonlinearity and DC-measured parameters is explained.

Chapter 6. Measurement of the Large-Signal Propagation Delay of Single Transistors

This chapter deals with performance of transistors in the time-domain, where large-signals are applied as they would be in digital circuits. A composite figure-of-merit for single transistors is presented. After consideration of measurement of single transistors, the ring oscillator structure is explained as a means to measure large-signal propagation delay. Modifications of ring oscillators are used to make possible the measurement of undesired variations of propagation delay. Time-domain measurement techniques which make use of these structures to observe such variations are described.

Chapter 7. Measurement of the Transient Response of Transistors

This chapter discusses how to determine if a transistor has a different response in a short time-domain vs DC. The transient response is considered for large-signal swings which switch the transistor from off to on, or the opposite. Several measurement methods are described by which this transient vs DC difference can be evaluated. Both observation of full transient response to a step function, and response to a short transient pulse, are described.

Appendix. Measuring with Controlled Temperature

The appendix describes some of the issues which must be faced in applying the various measurement techniques to samples which are held at higher or lower temperature than room temperature. Though not directly concerned with measurement method, temperature control has implications for the safe and reliable execution of the measurements presented in the rest of the book.

Chapter 2
High-Frequency Test Equipment, Connections, and Contact with Transistors

Measurement of electrical properties requires stimulus and detection equipment. Equipment for high-frequency and short time-domain measurements can be quite sophisticated and specialized. Detection instruments are used to measure the signals coming from the transistor being evaluated. The measurements also require instrument which provide a time-dependent or frequency signal to *stimulate* the device, such as pulse generators and sine wave generators.

This chapter describes the function of three types of equipment used for measuring signals coming from the transistor: spectrum analyzer, vector network analyzer, and oscilloscope. It also discusses the important topic of high-bandwidth, high-fidelity signal propagation using coaxial cables and describes the microwave probes which are used to bring signals to and from the transistors being tested.

2.1 Introduction

Stimulus and detection equipment is needed to bring signals (in the broad sense, as defined in Chap. 1) to and from the transistor. In the topics covered in this book, those signals might be of high-frequency ("RF") or short time-domain, and the test equipment is fairly sophisticated and must have high-bandwidth. In addition to these instruments, there are also the very familiar DC instruments, such as power supplies, voltage sources and current sources, and the combined source and measurement units which force a current or apply a voltage and accurately measure the corresponding voltage or current. While the stimulus instruments are fairly simple, the capabilities, and limitations of the signal detection equipment must be well understood, and therefore these instruments are summarized briefly in this chapter, and their use is discussed in more detail, as needed, as they are deployed in measurements in subsequent chapters. Because of the high frequency of many of the signals in most of the situations described here, interconnections must be made using high-bandwidth coaxial cables, which are a type of transmission line, which

© Springer Nature Switzerland AG 2022
K. A. Jenkins, *RF and Time-domain Techniques for Evaluating Novel Semiconductor Transistors*, https://doi.org/10.1007/978-3-030-77775-3_2

must be properly used and understood. The connections of the signals transmitted through the coaxial cables are made with "microwave" (again, not a precise term) wafer probes, which maintain the high-bandwidth of the transmission lines, and which eliminate the need for mounting samples in a package.

2.2   Signal Detection Instruments

There are three principal instruments which will be used extensively in the measurement techniques described in the book. They are (a) the spectrum analyzer, (b) the vector network analyzer (VNA), and (c) the oscilloscope. As instrument manufacturers continually advance the capabilities of their equipment, some features of each of these instruments are being implemented on the others. For the purpose of understanding their functions, though, it is best to regard each instrument as quite distinct from the others. All of these instruments are used to visualize a signal in its relevant domain, time or frequency, and as such were originally based on cathode-ray tubes (CRTs) which displayed a measured and amplified signal by deflecting an electron beam striking a phosphorescent screen as the time or frequency was swept. Now all measurement instruments have a large digital processing in the signal chain, and the displays are generated by discrete digitization of the detected and amplified input signals. While this digital processing has led to vastly improved equipment, it can sometimes impose limitations of which the user should be aware. Sophisticated digital signal processing (DSP) makes it possible to switch between time- and frequency-domains using fast Fourier transforms (FFTs), but generally the performance of the single domain instrument is superior.

Most instruments these days are equipped with a computer interface which gives a means to control the operation of the equipment and to transfer its measurements to the computer. The predominant interface is the General Purpose Interface Bus (GPIB), which is a hardware standard and a signal protocol. Ethernet and USB are also becoming common. Some of the instruments described below are also available in compact USB-connected devices which do not have live, on-board displays of the measurement, and sometimes poorer performance than the conventional equivalents. Instead of a display, the computer which operates them, acquires the data, and displays the data on a computer. This makes it possible for very powerful data processing software, but it can sometimes limit the ease with which multiple instruments can be operated simultaneously, and sometimes lacks the live "feel" of a traditional instrument display.

2.2.1   Spectrum Analyzer

A spectrum analyzer is used to measure the rms *power* of an AC signal as a function of a frequency, *i.e.*, its power spectrum, usually just called its spectrum. Hence, it is a swept frequency instrument. As it measures just the magnitude of the power

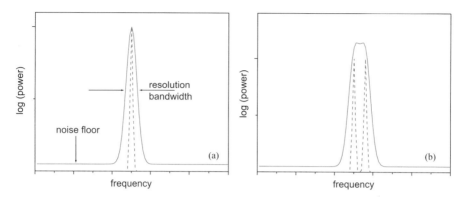

Fig. 2.1 Illustrations of frequency signals detected by a spectrum analyzer. The dashed lines indicate the actual signal, and the solid lines indicate what the spectrum analyzer displays. The width of the displayed signal is governed by the resolution bandwidth, and the noise floor represents the minimum signal which the spectrum analyzer can measure. (**a**) The input signal is a single pure frequency, (**b**) the input signal is two pure frequencies

without regard to phase, it makes a scalar measurement. Since the frequency is swept, it necessarily measures the time-averaged signal; that is, it cannot be used to make a one-shot snapshot of a signal's spectrum. Spectrum analyzers are free-running, that is, the measurements are not triggered by measurement signals and the sweeps run continuously although the start of the sweep can be triggered. (But see the comment on real-time oscilloscopes, below.) Some modern spectrum analyzers have a semblance of triggering by selecting certain frequencies to display, but they are still essentially free-running.

An illustration of the measurement of the spectrum of a single, almost pure, frequency, is shown in Fig. 2.1(a). The dashed curve shows the true signal spectrum, while the solid curve indicates what might be measured with the spectrum analyzer. Spectrum analyzers typically display the power on a logarithmic scale, in dBm, in order to display the huge range of power which they can detect. Note that dB indicates a dimensionless ratio,

$$dB = 10\log\left(\frac{P}{P_{ref}}\right) = 20\log\left(\frac{V}{V_{ref}}\right) \tag{2.1}$$

where the P and P_{ref} are two power levels, and V and V_{ref} their corresponding voltages. To express absolute power, a reference is specified, so that dBm indicates that the reference power, P_{ref}, is 1 mW. Using 1 mW as a reference can lead to confusion when converting from power to voltage in volts, and back. The unit dBW takes the watt as a reference power, which would not cause confusion, but this unit is never used in microelectronics instruments.

The spectrum analyzer has a 50 Ω input resistance,[1] and is usually connected with an intervening DC block, which is an inline capacitor which prevents a DC voltage level from damaging its sensitive receiver. The frequency range can be selected, as can its frequency resolution, which is often called the resolution bandwidth. The measurements can be averaged and filtered on the instrument. Figure 2.1 illustrates the effect of resolution: In Fig. 2.1(a), the measured spectrum is broader than the actual physical spectrum, but in Fig. 2.1(b) two closely spaced frequencies cannot be distinguished. This occurs if the resolution bandwidth is too coarse. An important parameter is the noise floor, which is the level above which a signal must be in order to be measured. Averaging, which is applied to the signal trace, can reduce the random fluctuations of the measured signal, but does not reduce the noise floor. But the resolution bandwidth also determines the equivalent noise bandwidth of a filter in the detector electronics chain which does affect the noise floor. The noise limit is determined by thermal (or Johnson) noise present in a resistor and is given by

$$P_n = kT\Delta f, \tag{2.2}$$

where P_n is the noise power, and Δf is the bandwidth of the noise signal. Reducing frequency resolution, which is equal to Δf, reduces the noise floor proportionally. It can also greatly increase the sweep time, as a smaller resolution naturally requires more time to cover the same frequency span. A good modern instrument can easily achieve a noise floor below −130 dBm, 130 dB less than 1 mW, which makes it an instrument of enormous dynamic range and sensitivity. (To put this in human scale, the difference between the sound of a jet engine, and the threshold of human hearing, such a pin dropping, is about 130 dB). In voltage terms, −130 dBm is 71 nV (7.1×10^{-10} V).

2.2.2 Vector Network Analyzer

A vector network analyzer, often called a VNA, or just network analyzer, also makes measurements in the frequency-domain, using swept frequency. Unlike the spectrum analyzer, it measures magnitude *and* phase of the received signal, hence the name vector. But phase measurement only makes sense only when there is a synchronized signal source for reference. Thus, unlike the spectrum analyzer and oscilloscope, the VNA generates the stimulus signal as well as detecting and measuring the device output. A typical VNA stimulates a terminal of a device, such as an input, and measures the signal reflected from the device, as well as its output from another terminal (its transmission). It also measures a fraction of the stimulating signal, to

---

[1] Some very low frequency spectrum analyzers, those operating below 1 Hz, have high impedance inputs, but they are not used in the work of this book.

provide a reference, and measures the magnitude and phase of each returning signal relative to the magnitude and phase of the input, using that reference. The phase is measured as the *difference* from the input phase and the magnitude of the signal is measured as the *ratio* to the stimulus signal. (Thus, a linear device shows the same VNA response, regardless of the stimulus power). It also works in the reverse direction, stimulating the output of the device, in order to measure the reverse transmission. The "test set" is a portion of the VNA which routes these signals. Similar to the spectrum analyzers, network analyzers are not triggered by device signals, but some very advanced models can be triggered to measure during a relatively short time pulse which is generated by an external pulse signal. Most VNAs use 50 Ω inputs, though 75 Ω units are also produced.

The measurement of phase is important for understanding reactive components of an electrical network. An example of measurements taken with a VNA is shown in Fig. 2.2. This is the measurement of a reflection coefficient, designated as S_{11}, of a device structure on a silicon wafer. Because a network analyzer measures two quantities, magnitude and phase, as a function of frequency, it is common to display them separately in rectilinear coordinates, as in Figs. 2.2(a, b). It can also display magnitude and phase in polar coordinates, as in Fig. 2.2(c), which uses real and imaginary coordinates instead of magnitude and phase. The frequency information is not labeled but can be read with a marker. In this example, the polar plot is more useful because the phase is related to impedance of a reactive device: to the experienced engineer, the data curve in Fig. 2.2(c) immediately suggests that the device being measured can be approximated by the schematic shown in Fig. 2.3. In this polar form, the measured magnitude of the signal (the radius in Fig. 2.2(c)) is displayed on a linear voltage ratio scale, whereas when it is displayed in rectilinear form of logarithm of magnitude versus frequency, as in Fig. 2.2(a), it is conventionally displayed on a *power* ratio scale as equal to $20 \log(S_{ij})$. Since it is a relative measurement, the units are dB, not dBm. Measurement data transferred to a computer, however, are usually in linear form. The stimulus signal is specified in rms dBm.

The noise floor of a typical VNA is somewhat higher than that of a spectrum analyzer although it can be maximized by reducing the bandwidth of the IF (intermediate frequency) stage of the detector, at a cost of longer measurement time, similar to the use of resolution bandwidth with the spectrum analyzer. Noise floor is not the only concern for a VNA: even a large signal can be noisy, but noise on large signals can be reduced by averaging, which may take less time to measure than a low IF measurement. It also should be noted that for some kinds of measurements, such as reflection coefficients, a VNA is most accurate when the impedance of the device which is being measured has a value close to 50 Ω. This is not due to the instrument sensitivity itself, but due to the nature of reflections from transmission lines.

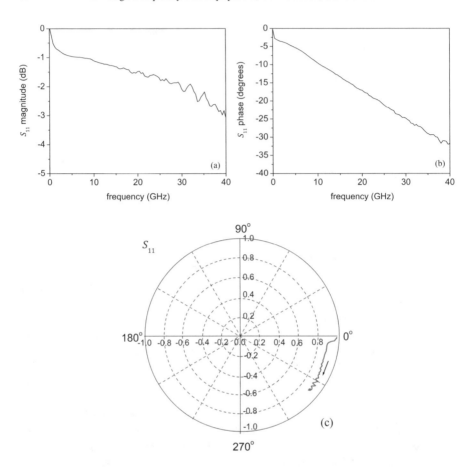

Fig. 2.2 A measured reflection coefficient (S_{11}) signal measured with a VNA. (**a**) Rectilinear plot of log(magnitude), in dB, *vs* frequency, (**b**) rectilinear plot of phase, in degrees, *vs* frequency, (**c**) polar plot of real and imaginary components of S_{11} with the real component on the horizontal axis, and the imaginary component on the vertical axis

2.2.3 Oscilloscope

The oscilloscope is probably the most well-known visualization instrument in electronics labs. It displays the voltage applied to its input as a function of time, i.e., a voltage waveform. This is particularly useful for measuring timing differences between signals, and for measuring the qualities, such as amplitude, rise-time, overshoot, and flatness, of a digital signal such as a voltage step or pulse. Of course, it can display a frequency-domain signal as a sine wave in time-domain. It can also display timing jitter of repetitive signals. Oscilloscopes can have input impedances of 1 MΩ or 50 Ω, but all of the oscilloscopes which have high-bandwidth and short time-domain sweep times, which are used in the studies in this book, have 50 Ω inputs. Unlike the spectrum analyzer and network analyzer, many oscilloscopes can measure

Fig. 2.3 Schematic of an
R-C network. This is a
network which
approximately results in
the measured signal in
Fig. 2.2

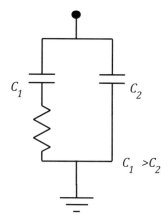

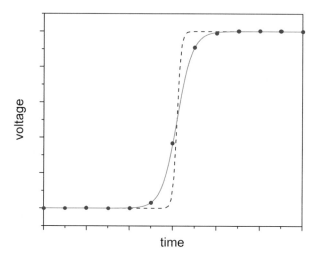

Fig. 2.4 Illustration of a time-domain voltage step measured with an oscilloscope. The dashed line indicates the actual signal, and the solid line indicates what the oscilloscope displays. The actual measurements are indicated by the dots, which the solid line connects

at frequencies down to DC. Some also have selectable AC or DC inputs. Selecting AC makes it possible to block the DC voltage and display the AC component of the voltage with high sensitivity. Oscilloscopes of highest bandwidth, however, usually have only DC inputs, but a high-bandwidth inline DC block, which is high-quality capacitor, can be placed between the signal and the oscilloscope input, if needed.

A typical oscilloscope trace is illustrated in Fig. 2.4. The true input voltage step applied to the input of the oscilloscope is indicated by the dashed line. The oscilloscope displays, however, the solid line. The increase of apparent rise-time of the applied voltage signal is due to the finite bandwidth of the oscilloscope's input amplifier, which, in this case, is lower than the maximum frequency components of the voltage step. It should also be noted that modern oscilloscopes are digitized, so that the measurements actually consist of a series of voltage-time dots, as shown in Fig. 2.4. The screen can display the dots alone or can interpolate between them to produce a continuous-*looking* waveform, and the interpolation can be linear or use a smoothing algorithm. If points are off the screen because the voltage sensitivity (volts/division) is too small or if there is a voltage offset, the interpolation may not work properly. The point density per unit time is generally engineered so as to not be a limit to the bandwidth, but the user should always be aware that the apparently smooth waveform is based on a set of discrete measurements. Oscilloscopes are usually described by the analog bandwidth of the input amplifier. As the voltage is digitized by an analog-to-digital converter (ADC), the voltages shown also have discrete steps, but since the ADCs typically have 10 bits or more, the voltage steps are essentially unseen.

High-bandwidth oscilloscopes can be either sampling or real-time types. Oscilloscopes are fundamentally triggered instruments, where a trigger initiates the capture of the voltage waveform, although they can be made to free-run, if desired. An input signal, which might be the same as the signal to be measured, or might be some other time-related signal, is the source of the trigger. Real-time oscilloscopes are the digital equivalent of a CRT oscilloscope: using fast ADCs, they digitize an entire waveform at once, making it possible to measure a one-time event (a "one-shot" trace). Compared to sampling oscilloscopes, real-time oscilloscopes have lower input bandwidth and lower point densities, but their performance is being continually improved. The sampling rate of real-time oscilloscopes, which specifies how many digitizations are made per unit time (usually expressed as billions of samples per second, or GSa/s), is the same as the point density. The best real-time oscilloscopes have sampling rates up to 250 GSa/s (4 ps between points). A one-shot oscilloscope trace can be used to construct a triggered power spectrum with a Fourier transform.

Sampling oscilloscopes operate by a different means. A *sample* of the waveform voltage is taken at a series of discrete time steps, and the waveform is constructed from the resulting samples. For this reason, they are sometimes called equivalent time oscilloscopes. The sample window can be very small to achieve high-bandwidth, and the sampled voltage itself is amplified and digitized with lower bandwidth electronics. (This is a form of the sub-sampling technique used in many circuits.) Thus, the sampling *rate* is quite low, even though the point density is generally very high, with intervals between points on the screen as small as 50 fs. Sampling oscilloscopes have the greatest bandwidth and sampling rate, and are thus used for the most demanding measurements and are particularly useful in communications work. However, because they sample the signal multiple times, they work only with a repetitive signal, just like the spectrum analyzer and network analyzer, and they also require a trigger signal. Sampling oscilloscopes have limited input voltage range, and if the input signals are large, it is necessary to interpose attenuators to protect the samplers from damage. Real-time oscilloscopes usually tolerate larger voltage levels. Noise floors for modern oscilloscopes are typically somewhat less than 1 mV (−47 dBm) and can be reduced by averaging. Some sampling oscilloscopes are equipped with modules for time-domain reflectometry (TDR) which is a method to measure cable lengths, imperfections in cables, and impedance of cable terminations.

2.3 Stimulus Instruments

For the purposes of nomenclature, a brief list of stimulus equipment is presented here.

– Signal generator: also known as a synthesizer or frequency synthesizer, it produces CW (continuous -wave) or pulsed pure sinusoidal waves, and often has the capability of applying frequency modulation (FM), amplitude modulation (AM), and phase modulation (ΦM) to the CW signal.
– Pulse generator: produces approximately rectangular pulses, that is, sharply rising and falling edges with a constant voltage between, with programmable high and low levels, pulse width/duty cycle, and period. They can usually produce a programmable burst of pulses.
– Square wave generator or clock generator: produces only 50% duty cycle pulses, with programmable frequency and high and low voltage levels.
– Data generator: produces programmable patterns of digital pulses, or data bits (1 s and 0 s) where the high and low levels, and the pulse period are programmable. They often include a pseudo-random bit sequence (PRBS) generator option to mimic the random data that might occur in digital communications.
– Waveform generator: can produce a variety of signals shapes on the same output channel, such as sine, triangle, sawtooth, square, pulse. They do not usually have as high a frequency as dedicated pulse or sine wave instruments.
– Arbitrary waveform generator: in addition to the functions of a waveform generator, it can produce repeating programmable signals of arbitrary shape by "connecting the dots" programmed into the generator's memory by user input.

2.4 Transmission Lines and Interconnections

The instruments used for high-frequency and short time-domain measurements all use coaxial cables for their signal connections. Coaxial cables used this way act as transmission lines, and serve to transmit, or transfer, electrical signals from one physical location to another. A rudimentary understanding of the properties of transmission lines is required to use many of the techniques described in this book. If one needs to apply a voltage (or current) to a distant device or apparatus, a wire is required to carry that signal, and an additional wire is needed for a return path or reference voltage level. Since the applied voltage cannot appear instantaneously at the other end, it propagates along the wire. While loose single wires can be used to apply voltages or currents, their inductance and capacitance limit the frequency with which the signal propagates.

Coaxial cables with BNC connectors are very commonly found in electronics labs. The BNC connector allows for a simultaneous quick connection of a voltage and its ground return, which would otherwise be connected with independent wires. In addition, the cable also maintains a ground reference for the signal. Even if there is no net current, as can happen when AC signals are applied, a reference is required for voltage to be defined, and the outer shield of a coaxial cable provides that reference, which is usually ground. This is one useful property of transmission lines. As the shield terminates the electric field lines originating at its center conductor, it also acts to shield the signal from creating or receiving electromagnetic signals from the neighboring environment, while providing an easy voltage and ground connection. A coaxial cable must be treated as a transmission line if the length of the cable is a significant fraction of the wavelength of the signal which is propagating. If it is of much smaller length, say, one percent or less of the wavelength, then the cable can be regarded as two wires with a lumped capacitance between them. Typically, the velocity of propagation in coaxial cable is about 0.7 times the speed of light, so from $v = \lambda f$, the wavelength in the cable of a 100 MHz sine wave is 210 cm; a 10 GHz sine wave, 2.1 cm.

Coaxial cable is not the only kind of transmission line. Another common structure, the coplanar wave guide, or CPW, is found on circuit boards and silicon circuits. Figure 2.5 illustrates these two types, but there are several others. The coaxial line, Fig. 2.5(a), has a center conductor, which is the signal line, surrounded by a dielectric and a ground, or shield conductor. The CPW, Fig. 2.5(b), has a center conductor, which carries the signal, with ground conductors on either side, and is essentially planar. It sits on top of an insulating substrate, which has a dielectric constant, and, of course, the air above the substrate is also a dielectric. As can be seen from Fig. 2.5, both transmission lines have signal and return wires in proximity. (The return is usually ground, but it isn't required.) The geometry of the wires is

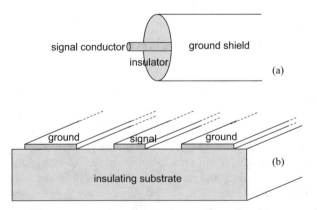

Fig. 2.5 Illustration of two common transmission line structures. (**a**) coaxial cable, (**b**) coplanar waveguide (CPW)

maintained throughout the length of the line, which can be thought of as infinite, so a transmission line can be regarded as defined by two dimensions. Energy flows through a transmission line as transverse electromagnetic (TEM) wave in the dielectric between the signal and ground wires.

If the transmission line is composed of zero-resistance metal, then it can be regarded as a ladder structure of distributed inductors and capacitors of infinitesimal length, as suggested by the drawing of Fig. 2.6(a), along which an electromagnetic wave travels [1, 2]. From this very simple concept, some of the important properties of transmission lines can be deduced. The most remarkable property of such a ladder structure is that, if it extends to infinity, its input impedance, that is, the impedance seen from the left in Fig. 2.6, is real, that is, a pure resistance. This important fact is often not fully appreciated upon first learning of it: an infinite ladder of only *reactive* elements has a purely *real* impedance. The impedance can be related to the reactances of the transmission line described by Fig. 2.6(a) by

$$Z_0 = \sqrt{\frac{L}{C}} \tag{2.3}$$

where Z_0 is called the characteristic impedance, and L and C are the inductance and capacitance per unit length, respectively. Transmission lines are also often called controlled impedance lines. What this impedance signifies is that when a voltage is applied to the input end of a transmission line, a current flows, too, governed by the relationship, $I = V/Z_0$. The resistive impedance for an *infinite* transmission is the same for both AC and DC activation. When the line is finite, this is no longer true, as there is no DC connection between the signal and its ground, and nowhere for DC

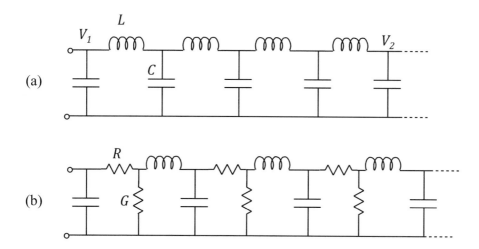

Fig. 2.6 Ladder representation of a transmission line. The ladder is assumed to go to infinity, and each step is vanishingly small. (**a**) A lossless transmission line, (**b**) a lossy transmission line with metal resistance and dielectric conductance

current to flow if the end of the transmission line is open. From Eq. (2.3), it is apparent that impedance is determined by the geometry and dielectric constant of the physical line, where the size of the center conductor and the separation of the conductor and ground(s) are the significant dimensions. In the case of coaxial cable, the commonly used impedances are 50 Ω and 75 Ω, and occasionally 92–95 Ω. (For comparison, free space has an impedance of 377 Ω.) As mentioned several times above, most high-frequency measurement instruments have input impedances of 50 Ω, precisely for the purpose of matching to the cable impedance, as will be discussed further below.

A voltage change presented to one end of a transmission line will propagate to the other end with a finite, albeit small, time delay. If the voltage is sinusoidal continuous wave, the phase of the wave will vary according to distance. Looking down the transmission line at different distances, say, from V_1 to V_2 in Fig. 2.6, it is obvious that a sinusoidal wave will have a phase difference which depends on that distance. Hence, a transmission line has a velocity associated with it, a phase velocity, which also depends on the geometry and dielectric constant:

$$v_p = \frac{1}{\sqrt{LC}} \tag{2.4}$$

In coaxial cable, this velocity is typically about 0.7 times the speed of light though special techniques to reduce capacitance can increase the speed. Finite lengths of coaxial cable are often referred to by the amount of propagation delay they introduce from one end to the other. Thus, a cable of slightly more than 40 inches might be called a 5 ns cable. In the frequency-domain, electrical time delay is sometimes expressed as phase change, or electrical length, per unit frequency. In this 5 ns example, a 200 MHz wave will be shifted by 2π radians from one end to the other, so its electrical length, β, is 2π radians at 200 MHz. The phase length of the cable, for arbitrary frequency, f, is $\beta = 2\pi f /200\ MHz$. More generally,

$$\beta = 2\pi f d \tag{2.5}$$

where d is the delay time of the cable. A coaxial component is often described as phase change/per frequency, rather than a time delay.

A non-ideal transmission line construction is illustrated in Fig. 2.6(b). This differs from the ideal line by including the realistic fact that the conductors have non-zero resistance (both DC resistance and skin effect), R, and that there is a possible conductance, G, from signal to ground in parallel with the capacitance, due to loss in the insulating dielectric, in which case, the transmission lines are considered lossy. These conducting paths do modify the impedance and phase velocity, but only very slightly and can be ignored for most purposes. However, it is important to note that they can attenuate the signal as it passes through the line, and the attenuation is worse at higher frequencies. Thus in a coaxial cable, a sine wave will suffer an amplitude attenuation, which depends on frequency, and a voltage step or pulse will be attenuated, and its rise-time will be reduced. The magnitude of these effects depends on the materials and construction of the cable, and on the connectors used, and it is good practice to be aware of such attenuation when measurements depend on accurate delivery and reception of signal transmitted through coaxial cables.

Of course, no transmission line is infinite and a finite line can behave quite differently. If the far end of the line is match terminated, that is, has a load, Z_L, which is a real resistance equal to the characteristic impedance, Z_0, of the transmission line, then the situation looks just like the infinite line: the current flows through Z_L and the input impedance is Z_0. Both AC and DC signals behave the same way. The current flows through the terminating resistor to ground as long as the voltage remains. This is the situation for measurement instruments with 50 Ω DC inputs. Driven by 50 Ω transmission line, the instruments receive the signal which was introduced at the input end of the line, so the line is merely transmitting that signal from one place to another, except for some attenuation. For a sine wave, this means that there is only a phase difference between the beginning of the line and its resistive termination.

The situation for a voltage pulse or step is illustrated in Fig. 2.7(a). A pulse or step introduced at the input end of a transmission will appear at the output end, the location of the load, delayed by the characteristic delay time of the line, T_0.

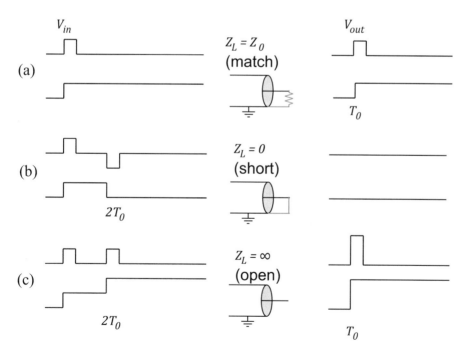

Fig. 2.7 Response of transmission lines to voltage pulse and voltage step for various termination conditions. The left shows the voltage *vs* time signals at the input to the line and the right shows the signals at the other end, the output, of the lines. The output is (**a**) loaded with a resistor equal to the characteristic impedance of the line, (**b**) loaded with a short, (**c**) loaded with an open (also called unterminated)

When the load is not a resistor equal to the characteristic impedance, a reflection must occur. Since $V = IZ_L$, but Z_L differs from Z_0, conservation of current requires that the reflected wave have a different voltage. Any number of loads can be imagined, and the voltages they create can be calculated. This is the basis of the technique known as TDR, or time-domain reflectometry, which uses reflected pulses to determine the distance and type of discontinuities (a place where the impedance differs from Z_0) in a transmission line. The *voltage* reflection coefficient, as distinct from the complex coefficient, S_{11}, is defined and determined by

$$\Gamma = \frac{V_r}{V_i} = \frac{Z_L - Z_0}{Z_L + Z_0} \tag{2.6}$$

where V_i is the input voltage and V_r is the reflected voltage. For most of the purposes of this book, only two resistive loads, other than the matched load, Z_0, need to be considered: the short and the open, for which the voltage reflection coefficients are -1 and $+1$, respectively. (From this equation, it can be seen that the relative sensitivity to load impedance of the reflection coefficient is greatest when the load impedance is close to Z_0, which is why the VNA is most accurate, as mentioned in Sect. 2.2.2 above.)

Using boundary conditions, the effect of these two terminations is easily determined. When the load is a short, zero ohms, the electrical field of the traveling wave must be zero at that point. Since the sum of the field of the incident wave and the reflected wave are added at that point, the reflected wave has its field, and therefore, its voltage reversed, and it returns back to the input inverted. This results in phase change of π radians for a sine wave, and the effect on pulses and steps is shown in Fig. 2.7(b). The input signals are the same pulse and step as in Fig. 2.7(a). Because the output is shorted, the output voltage remains at zero, but for the short pulse, an inverted pulse is reflected back toward the input. For the step, the input is eventually forced to zero also, but prior to that a pulse is created, equal in duration to the round-trip delay of the transmission line.

In the case of the open termination, where the load is infinite, at the end, there is no path for the current to flow to ground, and the reflected wave travels back to the source, creating a voltage of the same polarity as the incoming wave. Where the incoming and reflected waves superpose and add, then, the net voltage is double that of the input. As illustrated in Fig. 2.7(c), for a short pulse, the input will have two pulses separated by the round-trip delay. A step function will have a second step, with the voltage finally reaching twice the input value.

This doubling for an open termination is an important consideration when driving a terminal of a transistor being tested: if there is no $50\,\Omega$ resistance to ground in the measurement system near the input to the transistor, and the transistor has very high impedance, the voltage applied to the transistor will be twice the value indicated by the stimulus instrument, whether a signal generator or a pulse generator.

The discussion of termination above ignores the source of the input signal. The signal somehow gets launched into the transmission line at some distance and is reflected, inverted or not, or absorbed, by the termination. However, in connecting transistors to test equipment with coaxial cables, it is useful to consider the source of signals which enter the line. A voltage signal can be regarded as a current signal driving a load. Particularly in the case of pulse generators, this is the best way to understand what voltage will be present under DC and pulsed operation.

Unlike power supplies and SMUs (source and measurement units) most pulse and signal generators deliver a current to reach the desired voltage assuming they will drive a 50 Ω load, and the output voltage doubles if they drive an open load (such as an FET gate). Signal generators and pulse generators used in high-frequency applications always have internal "back termination," or "source termination" as shown conceptually in Fig. 2.8. The generator sends a current to an internal resistor and to the output of the instrument. The purpose of the source termination is to absorb any reflections returning from an unmatched load at the end of the transmission line so the reflection doesn't reflect at the instrument and travel back to the cable again. The current level is such as to drive the output to the desired voltage, under the assumption of a total resistance to ground of $Z_0/2$. (This is obviously not the situation if a transistor or circuit being tested is driving the transmission line.) If the transmission line is open at the far end then the voltage is doubled, and is maintained at this voltage, by the instrument, as long as the pulse lasts. Although this doubling is a transmission line effect, it occurs even when the physical line is much shorter than the wavelength, where transmission line effects can be ignored. In fact, it occurs at DC, as can easily be seen from Fig. 2.8. If, for example, a pulse generator is programmed to send a short 1.1 V pulse from a DC offset of −0.1 V, i.e., a −0.1 V to 1.0 V pulse, it will deliver a −0.2 V to 2.0 V pulse if the cable is connected to a device with high impedance. The DC level will remain at −0.2 V. An alternate way to send a pulse with a DC offset is to pass it through a bias tee described in Sect. 2.5.

Doubling of voltage may damage a transistor and can certainly lead to confusion and misleading results. Some modern pulse generators and waveform generators can be programmed for 50 Ω or "high-Z" output to adjust the actual applied current to achieve the desired voltage at the end of the cable, according to the cable's physical termination. But using 50 Ω termination consistently makes life easier.

Although an infinitely long transmission line has an input impedance which looks like a resistance, Z_0, when the line is finite, there are some circumstances in

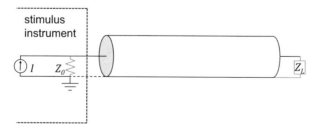

Fig. 2.8 Schematic illustration of the source termination used in equipment in which the output drives transmission lines and resistive loads

which the input impedance differs from Z_0. This happens when a CW (continuous wave) sine wave is applied to the input of the line. Unless the line is terminated in the characteristic impedance, just as in the voltage pulse and step described above, some of the signal is reflected back from the termination and, since the signal is continuous, the reflection interferes with the input wave at the beginning of the line. The sum of the input and reflection is a net wave of amplitude differing from the input wave, which means that the line has an effective impedance which differs from Z_0. Although this may seem like an annoyance, it is often deliberately used in microwave design for impedance transformation and matching. The effect can also be used to make resonators. But as coaxial cables are used extensively in connecting instruments and test devices, this effect can also find its way into measurement set-ups, where if not understood, it is an annoyance. But when understood, it can sometimes be used to advantage, as will be seen in Chap. 4 of this book.

The input impedance of a transmission line as a function of length, frequency, and termination can be computed from the general equations of transmission lines, but it is easy to see what happens for two special cases, called the quarter-wave and the half-wave lines, by examination of Fig. 2.9. In Fig. 2.9(a), a sine wave is introduced into the transmission line from the left. Assume the transmission line is lossless, so no amplitude is lost. If the end of the line is unterminated, i.e., open, the reflected wave, which travels from right to left, does not change sign. If λ is the wavelength and the physical length of the transmission line, l, is $\lambda/2$, a *half-wave line* (in frequency language, the electrical length of the line is π radians), the round-trip delay of the reflected signal is λ, so the reflection from an open termination causes the reflection to be in phase with the input, as shown in Fig. 2.9(a). When it arrives back at the input, the superposition of the input and reflection is a wave of

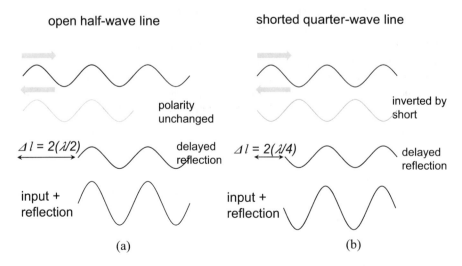

Fig. 2.9 Examples of the effect of termination on input impedance of transmission line stimulated with continuous wave sine wave. The impedance looks infinite for (**a**) an open-terminated half-wave line and (**b**) a shorted quarter-wave line

twice the input amplitude, so the input impedance is infinite, i.e., it acts like an open line. It can be shown, in the general case, that the input impedance of a half-wave line is the same as its load impedance. This open condition repeats for every $\lambda/2$ of transmission line length, e.g., $\lambda/2$, λ, $3\lambda/2$, but for other wavelengths, the impedance is less.

If the line is terminated by a short, as in Fig. 2.9(b), it will be inverted and reflected back to the input, as shown. If the physical length of the line, l, should happen to be $\lambda/4$, the *quarter-wave* case, the delay of twice the line's length means that the reflected wave will return to the input with a phase difference of $\lambda/2$, so the inverted wave will be in phase with the wave at the input. Hence, the total amplitude at the input of the line will double, which is the condition expected when its impedance is infinite, i.e., an open. In the opposite case, if the termination is an open, the returning wave will be exactly out of phase with the input, so the net voltage will be zero, i.e., a short. In the general case, it can be shown that the input impedance of a quarter-wave line is related to the reciprocal of its termination, Z_L, by $Z_{in} = Z_0^2$ /Z_L. Note, as indicated by the name, that this condition occurs when the line is a quarter of a wavelength. It is easy to see that the same result is also obtained for lines having lengths of $3\lambda/4$, $5\lambda/4$, etc., repeating every half-wavelength. These relationships are not perfect if the transmission lines are lossy, as all are, so the impedances are not exactly as predicted, but the alternation of high and low input impedance still occurs. For the usual case of a fixed physical length of transmission line, changing frequency changes the wavelength, and by doing so the signal can pass through $\lambda/4$ and $\lambda/2$ conditions, so the input impedance will swing between and Z_0^2/Z_L and Z_L, for example, between 0 and infinity, if the line is open-terminated ($Z_L = \infty$).

For a numerical example, consider a coaxial cable where the velocity of propagation, v, is about 0.7 times the speed of light, or 2.1×10^{10} cm/s. Then, from $v = \lambda f$, the wavelength in the cable is 21 cm/f [GHz]. A cable 21 cm long will have a quarter-wave condition for frequencies of 250 MHz, 750 MHz, 1.25 GHz, 1.75 GHz, 2.25 GHz, etc., and a half-wave condition for 500 MHz, 1GHz, 1.5 GHz, 2 GHz, etc.

Note that when a measurement instrument is driving a transmission line, the quarter-wave and half-wave effects do not occur, because instruments are source-terminated in the characteristic impedance they are designed to drive. Therefore, the reflected wave is terminated and absorbed when it reaches the input and does not interfere with the outgoing wave.

2.5   Cables, Connectors, and Bandwidth

The discussion of transmission line cables is incomplete without considering their bandwidth. Most of the techniques described in this book require the delivery or detection of high-frequency or short time-domain, i.e., high-bandwidth, which is a sloppy term meaning passing high-frequency signals. Cables introduce attenuation because of resistance and conductance through the dielectric. Cables also have strong cutoff frequencies according to their geometries. And there are many types of RF coaxial cables which vary in diameter, mechanical flexibility, and attenuation. The most common flexible cables are RG 174 and RG58 (RG is an old term meaning radio guide). Rigid, semi-rigid, and conformable cables are also used. However, bandwidth is usually determined not by the cable, if it is of decent quality, but by the connectors used with it. There are a variety of coaxial RF connectors commonly available, distinguished by size, ruggedness, connection method, and recommended maximum frequency, and most of them are available for use with a variety of coaxial cables. The recommended maximum frequency is about 90% of the connector cutoff frequency. Table 2.1 shows some of those most commonly used.

The numbers in parentheses are the equivalent rise- and fall-times due to the connector bandwidth. That is, an instantaneous rising or falling edge would be transmitted with this finite value. This uses the conventional formula which is based on a step applied to an RC filter:

$$t_r \cong \frac{0.35}{f_{3dB}} \left(10\% - 90\%\right) \tag{2.7}$$

where t_r is the rise-time, measured from the 10% to the 90% points of the waveform and f_{3dB} is the 3 dB bandwidth measured in ns and GHz, respectively. The 3 dB bandwidth, where the power is reduced by one half, is also known as the cutoff frequency of the filter. Cutoff frequency has a different meaning when applied to transistors, which can lead to confusion. The maximum frequency in the table above is *not* the 3 dB bandwidth, but it is close enough to give an idea of how short rise- and fall-times can be when passing through the various connectors.

Just as bandwidth determines how high a frequency can be supported, rise- and fall-time determines how short a time signal can be used: a cable or instrument which limits rise-time also limits the width of a pulse. It is worth being careful about what is meant by the bandwidth-rise-time equation, Eq. (2.7) above. The equation is based on stimulating a low-pass RC filter network with a sine wave or a voltage step. Analyzed in the frequency-domain, it has the response shown in Fig. 2.10(a). At sufficiently high-frequency, the voltage is attenuated. Specifically, at angular frequency $\omega = 1/RC$, the voltage magnitude is attenuated by 0.707; the power, 0.5,

Table 2.1 Common coaxial RF connectors

Coaxial connectors	Maximum frequency (rise-time)	Comments
BNC	4 GHz (88 ps)	Bayonet mount
SMB	4 GHz (88 ps)	Press-on
Type N	11 GHz (32 ps)	
SMA	18 GHz (19 ps) normal	25 GHz with modifications
3.5 mm	26.5 GHz (13 ps) normal	34 GHz theoretical
2.92 mm (K)	40 GHz (9 ps)	
2.4 mm	50 GHz (7 ps)	
1.85 mm (V)	60 GHz (6 ps)	

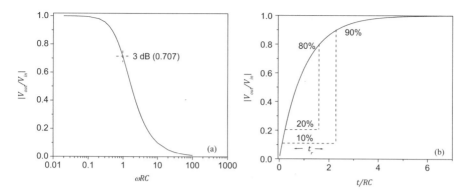

Fig. 2.10 Explanation of the correspondence between bandwidth and rise-time of an RC (low pass) filter network. (**a**) frequency-domain response, illustrating 3 dB bandwidth, (**b**) time-domain response illustrating rise-time

or 3 dB. The frequency at which this occurs is called the 3 dB bandwidth, or the cutoff frequency. In the time-domain, a positive voltage step causes the voltage to rise with an exponential shape, as shown in Fig. 2.10(b), where the exponent has the argument $-t/RC$. The rise-time is defined as the time it takes for the signal to rise from one voltage to another, typically 10–90% or 20–80% of the maximum. The 10–90% definition leads to the Eq. (2.7). The 20–80% definition leads, instead, to

$$t_r \cong \frac{0.22}{f_{3dB}} \; (20\% - 80\%).$$

(2.8)

Although the 10–90% formula is most commonly used, theoretically the 20–80% definition is equivalent. But from the point of view experimentally determining bandwidth by oscilloscope measurement, Eq. (2.8) is safer. Because the exponential function starts to level off, estimating a 90% voltage point is likely to lead to a large error. It is also to be noted that Eqs. (2.7) and (2.8) strictly refer to *RC* networks and should only be used as a useful guide when there are other factors which limit bandwidth.

All RF connectors except the BNC and SMB are screw-type connectors and should be tightened with a torque wrench which ensures good electrical contact but prevents damage from over-tightening. SMA, 3.5 mm and 2.92 connectors are mechanically compatible, but the lower manufacturing quality of the SMA can damage the precision 3.5 mm and 2.92 mm connectors if they are mated, and it is considered bad practice to screw them into each other. Rather, between-series adaptors, such as SMA to 2.92 mm, should be used to connect different RF connector types. 2.4 mm and 1.85 mm are mechanically compatible with each other, but not with the rest in this list. BNC and SMB connectors are seldom used for high-frequency work. BNC connectors are rugged and good for quick connecting and disconnecting. SMB connectors are very small and useful when there are a large number of connections needed in a small space. Type N connectors are usually just

found on instruments, where they are used for their exceptional ruggedness. K connectors or 2.4 mm connectors should be used for vector network analyzer measurements, but SMA connectors can be used for most of the other techniques discussed
in this book.

2.6   Bias Tees and Resistor Tees

Two seemingly minor components of many measurement systems are the bias tee
and the resistive tee. The bias tee, a simple device, illustrated in Fig. 2.11, separates
the DC bias level from the AC signal applied or received from the device. As shown
in the Fig. 2.11, as it works in both directions, it can add or remove the DC voltage
of the AC signal. The capacitor has a low impedance for AC and passes the signal
from left to right, or right to left, without much attenuation. A coaxial element, the
tee is inserted "inline" in the coaxial cable leading to or from the equipment. The
inductor, having high impedance for high-frequency, allows the DC voltage bias,
V_{bias}, to reach the transistor, while blocking the AC signal, and the capacitor blocks
the DC voltage in one direction. (The inductor in the power supply path can also be
used to suppress or reduce moderate frequency power supply noise.) Many models
can be obtained; the size of the capacitor and inductor affect the upper and lower
frequency over which the bias tee works. It is also to be noted that the resistance of
the inductor in the tee, typically a few ohms or less, can cause a voltage drop, reducing the voltage at the transistor terminal, if the current passing through it is large.
Vector network analyzers usually have bias tees built in to their test sets.

   Bias tees can also be used to level-shift short pulse signals, which might be
required if the instrument providing the signal does not have the range of amplitude
and offset required by the test. For example, suppose a novel device requires testing
with short pulses which reach 2.5 V, but the pulse generator cannot produce voltages
greater than 2.0 V. Then a bias tee can, in principle, be inserted to provide a 0.5 V
offset. However, this must be done carefully since a signal which doesn't have 50%
duty cycle results in a DC offset voltage when passed through a capacitor, where
duty cycle is a description of how often an alternating signal is at its high value. A
signal which alternates between high and low signals with the same duration is a

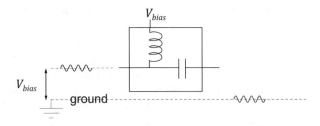

Fig. 2.11 Schematic of a bias tee and its function

50% duty cycle. A signal passing through a capacitor has zero *average* voltage, so a
pulse train with a duty cycle of D (defined as pulse width divided by pulse period),
and height H, has its low level, L, shifted according to

$$\frac{L}{H} = \frac{D}{1-D} \tag{2.9}$$

$$L = DA$$

where $A = (H + L)$ is the pulse amplitude as defined in the voltage waveforms
shown in Fig. 2.12. In Fig. 2.12(a), the 50% duty cycle signal swings around ground
(L and H are equal), but in (b), a short pulse of low duty cycle passes with almost no
change of the low level. The pulse is passed as shown if its duration is much smaller
than the time constant corresponding to the low frequency cutoff of the tee. In
Fig. 2.12(c), the baseline is shifted considerably as the duty cycle gets large.
Therefore, to control the offset level of a pulse with a bias tee requires adjusting the
DC voltage applied to the tee knowing the offset already caused by the duty cycle
of the pulse.

When a bias tee is used to offset a pulse, the doubling which occurs when a
coaxial cable is not terminated takes a different form from the description in Sect.
2.3. The DC level of the pulse is blocked by the tee, and the offset is provided from
the V_{bias} control. Thus, the pulse will double, but the offset will not. Using the example of Sect. 2.3, if the offset is −0.1 V and the pulse is 1.0 V, for a net pulse of 1.1 V,
using a bias tee, the doubling results in a pulse which swings from −0.1 V to 2.0 V.

The resistor tee is another simple device for dividing the signal into equal or
unequal parts using a resistive division network. The resistors are mounted in a carefully designed mechanical structure which attempts to preserve the controlled
impedance of a transmission line. The most common form of a resistive tee is the
three-way symmetric power splitter, or power divider, shown in Fig. 2.13. This
device takes a signal in any one of its three inputs and divides it equally to the other

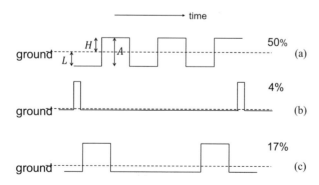

Fig. 2.12 Illustrating how duty cycle of voltage pulses passing through a capacitor can cause a
voltage shift of the pulses. (**a**) A 50% duty cycle results in the pulses swinging about zero, (**b**) a 4%
duty cycle pulse passes the pulse with almost no level shifting, (**c**) a 17% duty cycle causes a significant offset

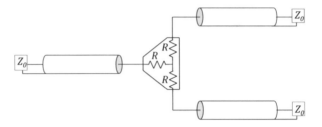

Fig. 2.13 Schematic of a resistive tee network

two branches. The output of the tees is meant to be terminated in the characteristic impedance, Z_0. When this is the case, it is easy to see that the resistors in the symmetric tee are equal to $Z_0/3$. Higher dimensional splitters are common, but not used in any the studies in this book. Although it is possible to use a resistive tee without terminating the output in the characteristic impedance, the reflection resulting from the unterminated output will combine again when entering the tee and will add to the signal at the terminated output, causing an unwanted waveform. Symmetric tees are often called splitter/combiners because two input signals can be added to form one output, but it is important to note that unless the electrical lengths of the input cables are perfectly matched, the signals will be summed with different phases or delays, resulting in unpredictable output. Asymmetric resistive tees can also be built, where the input and output assignments are fixed. Also available are taps, or pickoff tees, where a small fraction, less than 1/2, of the input is present at one of the outputs. These, of course, are asymmetric tees. Typical available fractions are 1/5 and 1/10.

2.7 Transistor Probing

In the illustrations of various measurement systems described in the book, the instruments are shown contacting the terminals of the transistors being tested. Because of the need for high-bandwidth signal transmission, this contact of the transistors best done using "microwave" probing of a wafer. While it is possible to mount a test device in a package which has appropriate connectors, it is challenging to find a package with high-bandwidth suitable for the evaluations described here. Also, packaging the device requires wire-bonding, which can lower the bandwidth and because of the damage to the bond pads, makes reuse of the device difficult. Probes, in contrast, require almost no sample preparation and may cause only minimal damage to the probe pads. Many devices on a sample wafer can be probed just by moving the sample or the probes, and samples can be easily changed. For these reasons, DC probes are widely used in industry and in experiments, and high-frequency equivalents are manufactured by several vendors. Microwave probes essentially extend the transmission lines all the way to the transistor contact pads, by transforming coaxial cable to coplanar waveguides which connect to the wafer with short probes. The invention of microwave probes, though not often recognized, was a major technological advance in the realm of transistor testing, giving engineers the ability to easily measure devices and circuits with high-frequency signals.

Figure 2.14 shows photographs of two examples of microwave probes from different manufacturers [3, 4]. Inside the body of the probe a transition is made from the coaxial form of transmission line into a coplanar waveguide structure while maintaining the controlled impedance. Typically, a microwave probe has three contacts to the transistor, two grounds and a signal, known as GSG, for ground-signal-ground. Other configurations, such as SG, GSSG, SGS, and GSGSG, can be purchased. In order to control impedance and maintain high-bandwidth, the spacing between the contacts is fixed, and, unlike DC needle-type probes which can be individually positioned with manipulators, these must match the probe pads of the transistor being tested. The bandwidth of microwave probes is usually dominated by the coaxial connector. Typically, the K-connector is the lowest frequency model and has a bandwidth of 40 GHz. As these probes use coaxial cables, which can sometimes be stiff and heavy, the probes need to be mounted on strong positioners which can withstand the strain of the cables. It is also possible to have probes mounted on a probe card which can make a more mechanically rigid measurement system, but this can only be used for probe pads which match the spacing of the probes. It is possible to have probes built where there is a 50 Ω resistor from signal to ground attached right on the probes [4], so that the problem of reflection from an unterminated line can be eliminated.

Fig. 2.14 Photographs of two kinds of microwave probes Photographs courtesy of FormFactor, Inc. [3] (left) and Picoprobe by GGB Industries, Inc. [4] (right)

2.8 Probe Pad Design

A typical layout of a transistor which is used in high-frequency measurements is shown in Fig. 2.15. Figure 2.15(a) shows the probe pad structure surrounding a hypothetical transistor device. The GSG arrangement of the probes corresponds, in this case, to source-gate-source terminals on the input and source-drain-source on the output. Note that the grounds of the input and output probes are connected by the source pads. This is important to maintain high-frequency signal transmission: the transistor can be regarded as something inserted in the middle of a transmission line going from input signal to output detector, where the signal flows without any perturbation due to discontinuity in the transmission line. Figure 2.15(b) shows a photograph, taken through the eyepiece of a microscope on a probe station, of microwave probes in contact with a test transistor. Although the details of the pad design in this example differ from the schematic drawing in Fig. 2.15(a), the concept is similar. An alternate pad structure which occupies much less space is discussed in Chap. 7.

A significant issue which must be faced in using microwave probes is the material used for the probe pads. Unlike DC needle probes, microwave probes tend to "scrub" the pads with substantially larger tips. The probe-to-pad contact area can be as large as 25 μm in diameter. Such a large contact area can result in removing material if the probe metal is very thin or if the metal does not adhere well to the substrate. Of course, scraping off metal will result in poor electrical contact. The commercial silicon processes provide a last metal level which is thick aluminum, generally intended for wire-bonding. This, or an equivalent thickness of gold, makes for a very good contact with microwave probes. But for exploratory transistors, fabricating this metal level may not be available. Generally, the pad metal should be softer than the probe tips, so they can dig into the pad to make a low resistance contact. There are several kinds of probe tips possible, such as tungsten, beryllium-copper, and nickel, each with different hardness. It is worth considering these options in combination with the available metals for pad fabrication, and their

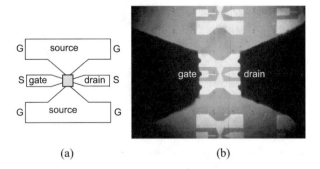

(a)   (b)

Fig. 2.15 Illustration of probe pads used for microwave probing of transistors. (**a**) Typical pad geometry, showing connections to FET terminals. (**b**) Photograph of microwave probes in contact with an FET with suitable probe pads as seen through the probe station microscope

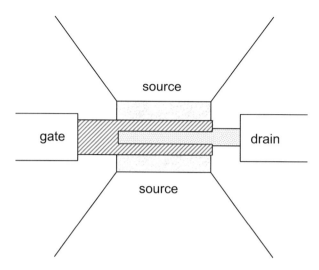

Fig. 2.16 Simple layout of a transistor to make contact with the probe pads of Fig. 2.15(**a**)

thickness. A good probe-to-pad contact should have a resistance of an ohm or less. Significantly higher resistances can be tolerated for low-current devices since they would not generate a large voltage drop. However, a continuous assured contact of both signal and ground is essential for high-frequency measurements. An intermittent loss of contact of either will lead to useless results.

Finally, the connection between the transistor and the probe pads must be optimized to minimize long interconnections, which otherwise lead to inductance, to excess capacitance, and to large resistance to the transistor. An example of a possible layout is shown in Fig. 2.16. This concept results in a two-finger device which takes advantage of the two source contacts of this double GSG layout. If the device is fabricated in an advance processing technology, where multiple metal layers are available, then multi-fingered devices can be constructed. On the other hand, if only a very simple technology process is available, a single finger gate may be required, resulting in one unused source contact. An example of this layout will be shown in the discussion of carbon nanotube transistors in Chap. 3.

References

1. R.Matick, *Transmission Lines for Digital and Communications Networks*, New York: Mc-Graw-Hill, 1969.
2. D. M. Pozar, *Microwave Engineering, 2nd Edition*, New York: John Wiley & Sons, 1998.
3. FormFactor, Inc., Livermore, CA; www.formfactor.com
4. GGB Industries, Inc., Naples, FL; www.ggb.com

Chapter 3
Measurement of the Frequency Response of Transistors

Determining the frequency response is a usual first step in assessing the performance of a novel transistor. Frequency response is established by measuring the gain of the device as a function of the frequency of its input signal and if possible, defining a frequency figure-of-merit where the gain has dropped by a certain amount or reached a fixed number. For conventional transistors, this is done through measurements of S-parameters using a vector network analyzer, and analyzing the measurements from the perspective of a small-signal model of the transistor, to obtain cutoff frequency and other frequency figures-of-merit. This technique is well-known and is common practice, but requires that the transistors be of sufficient size to develop a strong signal in the network analyzer. The technique may not work well, or at all, with novel transistors which have very low currents and a gate capacitance which is small compared to the probe pad capacitance.

In this chapter, the complicated S-parameter measurement method is explained in detail, including the important topic of de-embedding. Alternate measurement methods are described for use with novel transistors when their output current is too low to use the S-parameter method.

3.1 Introduction

After the initial trial of a new transistor which yields good DC characteristics, it is natural to want to know if it has good frequency response. This can mean several different things, such as (a) will it make a fast digital switch? (b) will it be a good high-frequency (RF) amplifier? (c) will it make a good mixer? It is necessary that any useful transistor will exhibit some kind of gain. That is, it will amplify its input, whether it is current, or voltage, or power. Although it is conceptually possible to build circuits by connecting devices having unity gain, as a practical matter no real circuit which performs a useful function has ever been demonstrated with unity-gain devices. Briefly, gain is needed in order to amplify the input so one transistor

© Springer Nature Switzerland AG 2022
K. A. Jenkins, *RF and Time-domain Techniques for Evaluating Novel Semiconductor Transistors*, https://doi.org/10.1007/978-3-030-77775-3_3

can drive a larger transistor, or multiple transistors, which enables interconnecting devices to make circuits of increasing complexity. So, in evaluating frequency response of a transistor, what is under consideration is how does its gain change with frequency? At what frequency does the gain become too low to be useful? In the time-domain, this can be viewed as how long it takes for the output of the transistor to respond to a changing input.

Silicon MOSFETs are good at many functions, which is why they are ubiquitous, but a new transistor may be better for one use or another, but not all uses. Graphene, as an example, was a new material (but not a semiconductor!) with very high mobility, with which transistors were made. Because their channel conduction could be modulated, but not turned off, some researchers had high hopes that graphene FETs would be good for very high-frequency analog circuits, but their use as digital switching transistors was never seriously considered. In the earliest development of a novel transistor, the analog gain as a function of frequency is the first performance property to be measured, and that is considered here. It is noted that in the measurement of frequency response, it is assumed that the transistor is stable. This is by no means guaranteed for a novel transistor, and the question of stability is taken up in later chapters in the book.

3.2 Methods for Low-Current Transistors

It might seem that to establish a new transistor's frequency response just requires applying a sine wave to the gate and measuring the output drain current, and increasing the frequency until the output current amplitude starts to diminish. However, a complete characterization of frequency dependence must come eventually from S-parameter measurements, which will be described subsequently in this chapter. The ultimate purpose of such measurements is the establishment of a mathematical model to describe the AC behavior, just as DC measurements are used for a SPICE model. The establishment of a mathematical model is absolutely necessary to design integrated circuits using the novel transistors (but the model must be appropriate for the type of circuit being designed). But in the early stages of a new device, such measurements may not be possible. S-parameters are measured with a vector network analyzer, which requires a significant drive current in its 50 Ω inputs in order to produce a power signal large enough to be well above the noise floor. And inasmuch as S-parameter measurements require subtraction of signals, as will be discussed later, the need for low-noise signals is very important. It's quite common to find that a new device, which is constrained by lithography and the amount of good quality channel material, has rather small drive current. For example, early carbon nanotube transistors had maximum DC drain currents (rail-to-rail) of about 1 μA [1]. Using one-tenth of that as a guess at the small-signal regime leads to a power signal of −93 dBm or less, dropping off which frequency, which is difficult to distinguish from noise in some VNAs (vector network analyzers). In addition, such transistors typically have small intrinsic gate capacitance so the frequency response is determined in part by the capacitance of the probe pads. Although this can be dealt with using the de-embedding technique which will be described later, in Sect. 3.3.4, it leads to significant errors when the pad capacitance is dominant.

In this situation, it is desired, instead, to do some direct or indirect test of gain *versus* frequency in order to establish that the frequency is not far below expectations. (For example, with carbon nanotube FETs again, some predictions were for operation into the THz regime [2]).

If the new transistor results from a modification of an existing transistor technology, such as an additional implant, a change in process, or a gate metal change, it is probably already suitable for high-frequency measurements. But a new transistor, such as one made from a two-dimensional material, may be initially suitable only for DC testing. The difficulty is that in order to do any high-frequency tests, input capacitance of the device must be appropriately small. A capacitive load at the end of a 50 Ω transmission line results in an *RC* time constant which attenuates the input signal, where the resistance is Z_0 (usually 50 Ω) and C is the capacitive load, which is the gate capacitance in this case. The corresponding corner frequency is $\omega = 2\pi f = 1/RC$, above which frequency the signal diminishes rapidly which masquerades as a drop off of the transistor output.

3.2.1   Transistor Structures

It is assumed here that new transistors are of the planar FET type. Figure 3.1 shows two constructions of planar FETs. In Fig. 3.1(a), the channel material is placed on a gate dielectric which has been deposited or grown on a low-resistivity substrate, such as doped Si. The deposition of the channel material can be done by sophisticated semiconductor processing techniques, or, as is often found in exploratory labs, a small piece of the material is mechanically placed on a substrate. Source and drain terminals are deposited on the material, perhaps customized to match the small piece of material, and the bottom of the substrate is metallized to form the

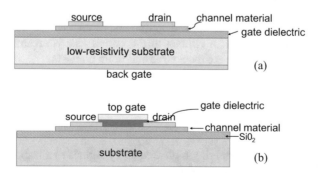

Fig. 3.1 Simple structures to make FETs with new channel material. (**a**) back-gated transistor, (**b**) top-gated transistor

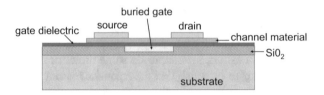

Fig. 3.2 Buried gate structure to make FETs with new channel material

gate electrode. This back-gated FET is very easy to make and is good for DC measurements, but is impractical for direct high-frequency operation because the gate electrode due to its large area (the entire substrate) has very large capacitance, which severely limits the frequency which can be applied to it. (The role of gate capacitance becomes explicit in the hybrid-pi model, which is commonly used to describe small-signal frequency operation, and is presented later in the chapter.) This construction also exposes the channel material to the environment, making it susceptible to changes due to surface contamination, unless a passivation process is applied. Thus, a top-gated transistor is generally preferable. Figure 3.1(b) suggests one top-gated structure, where now the physical area of the gate, rather than the area of the entire substrate, determines the gate capacitance, and the channel is better protected from the environment in all directions. Unlike the back-gated structure, the substrate is either insulating or semi-insulating and covered with a dielectric layer to prevent conduction from source to drain, but also reduces unwanted AC signal propagation through the substrate. Such a top-gated structure is, of course, necessary for use of the transistors in integrated circuits and is found in all successful commercial device technologies.

A more sophisticated, but attractive, fabrication technique uses a buried gate structure to build a transistor with mechanically deposited channel material [3]. Illustrated in Fig. 3.2, it requires significant semiconductor processing techniques, such as etching, filling, planarizing and via construction, but offers small gate electrodes, reduced sensitivity to environment, and thin dielectric films. As shown in the figure, gate electrodes are formed by patterned etching of shallow trenches in an insulating layer, which are then filled with an appropriate metal, such as tungsten. This is followed by deposition of a gate dielectric. The channel material is then placed or deposited on the surface, and source and drain electrodes deposited, just as in the back-gate structure. A contact to the buried gate is created so it can be probed on the top surface, as are the source and drain contacts.

3.2.2 Rectification Technique

In spite of the problem of applying high-frequency signals to a back-gated transistor, it has been shown in a clever experiment, that a back-gated FET can be used to examine frequency response [4]. In this experiment, the frequency response of a

p-type carbon nanotube FET (CNFET), where a single gated nanotube connecting source and drain, was studied. Recognizing the difficulty of direct measurement of small-signals because of the back-gate capacitance, this experiment made use of an RF in/DC out technique to eliminate the problem of measuring tiny output signals. Instead of measuring the small RF signal coming from the CNFET, the transistor is connected so that a small average DC signal can be measured. Precise measurement of small DC currents is quite easy, compared to measuring corresponding RF signals. (The general idea of using AC signal to stimulate the device, and DC current measurement to detect its output can be a powerful technique. It is used again in a very different application in Chap. 7.) The transistor is connected in the diode-connected configuration shown in Fig. 3.3, with the gate connected to the drain which makes a three terminal transistor act like a two-terminal diode. All of the connections are made with coaxial cables. Instead of applying the input to the gate, it is applied to the source using a probe pad, and the gate and drain are connected and the DC drain current is measured. This method avoids the problem of driving the large back-gate. Microwave probes are used to contact the FET source. The device construction was similar to the back-gated device shown in Fig. 3.1(a), except that the gate dielectric (SiO_2) thickness was thinned locally so similar gate and drain voltages could be used. The gate-to-drain connection leads to the transistor acting as a rectifying diode between the source and drain, as any non-linear device will act to rectify the AC input. Even without knowing the shape of the output characteristic, the existence of a threshold voltage is sufficient for rectification. Furthermore, no net DC signal is generated by any of the input sinusoidal signal passing from input to output through capacitance from probe pads to the substrate, as signals passed by capacitance average to zero. Compared to the direct power gain technique described in the next section, the elimination of parasitic signal is quite advantageous. However, it must be noted that while the large gate capacitance is not a factor in this back-gated configuration, the capacitance of the source probe pad to substrate does significantly attenuate the RF input, and considerable effort is needed to account for this attenuation in the experiment.

Note that for a diode, or diode-connected FET, to act as a rectifier, the applied signal must be large; otherwise, by definition, the device acts as a linear network. This is demonstrated in a representation of the measurement shown in Fig. 3.4(a). This figure shows the measured drain current as a function of the bias applied to the gate and drain, that is, the diode bias. The signal applied to the source is of a low enough frequency, 1 MHz, that it is not attenuated by the RC effect as described

Fig. 3.3 System used to measure frequency response using rectification method [4]. The connections use coaxial cables

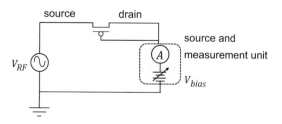

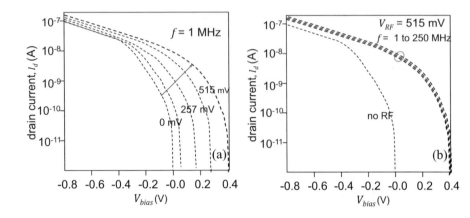

Fig. 3.4 Measured frequency response of carbon nanotube pFET using the rectification method [4]. (**a**) Drain current vs bias voltage at low frequency, for various RF voltage levels, (**b**) drain current vs bias voltage for large RF voltage, for various frequencies

above. The curves show the drain current as a function of bias, measured for various amplitudes of the RF signal. When the RF signal is 0 mV, the curve is just the DC transfer characteristic, showing a threshold voltage of about −0.3 V. The shape is determined by the configuration where gain and drain voltages are the same and differs from the usual output curves where the drain voltage is a constant. When the amplitude of the RF signal is increased, the DC current increases for regions where the bias voltage is greater than about −0.4 V. This is the result of the transistor having a threshold voltage: when the bias is high enough that the RF signal swings from the exponential sub-threshold characteristic to the "on" region, the rectified current increases with increasing amplitude of the RF signal.

To measure the carbon nanotube FET, frequency response is a matter of increasing the RF frequency which is applied to the source and measuring the DC current. Although high-frequency microwave probes were used, the parasitic source pad capacitance limits the frequency which can be achieved with this setup. Because this input capacitance is fairly large, the corner frequency, where the amplitude drops by 50% of the input signal is about 300 MHz, and it is necessary to calibrate the signal and adjust its amplitude to be equal for all the frequencies used. Having done this, data looking like Fig. 3.4(b) are obtained. This figure shows the current measured with a large (515 mV) signal amplitude applied at frequencies from 1 MHz to 250 MHz, the maximum which can be applied based on the calibration. In contrast to the case of no RF signal applied, the same current is measured for all frequencies. This implies that the carbon nanotube FET has no diminution of frequency response up to 250 MHz, which, in turn implies that the fundamental frequency response must be much higher than the upper limit of this experiment system. Although this measurement method of the experiment was used to solve the problem of a large capacitance of a back-gated FET, it can equally well be used for a top-gated transistor, though direct measurements, as described next, may seem more appealing.

Although this measurement technique clearly implies that the carbon nanotube FET can operate at high frequency, it does not give very direct, visible, measure of the ET's response to a high-frequency input since the output signal is not seen except through the indirect rectification into a DC current.

3.2.3   Direct Gain Measurement Technique

A top-gated (or buried gate with a top contact) transistor structure has more measurement possibilities: high-frequency signals can be applied directly to the gate, unattenuated by the large capacitance of the substrate, to determine to what frequency the drain RF output follows the applied input. Because the transistor is driving a 50 Ω load and yet has a much higher output impedance, this kind of measurement does not lead to a well-defined figure-of-merit, such as transducer gain, but it can be useful as a very simple and easy to implement first direct assessment of the ability of the transistor to operate at high frequency.

The measurement consists of driving the gate of the transistor with a signal generator, and observing its drain signal with a spectrum analyzer, as illustrated in Fig. 3.5. DC voltages are applied, using bias tees, to both gate and drain, and the device is probed with typical microwave probes as described in Chap. 2. The input frequency *is* swept and the output power at the same frequency is measured. A spectrum analyzer is preferred over a power meter because of its much greater dynamic range and the ability to measure the power of a single frequency. Since the output is loaded with the 50 Ω input of the spectrum analyzer, a low-current device will appear to have power *loss*, rather than a gain.[1] This is just a consequence of the low input impedance of the spectrum analyzer and the low output current from the transistor and has nothing to do with frequency response. However, the signal must be large enough to measure. Using the hybrid-pi model of a transistor, which is described later in the chapter and in Chap. 4, it can be shown that, at DC or low frequency, that is, ignoring intrinsic and external capacitances, the observed gain is just the ratio of power out to power in,

Fig. 3.5 System used for direct measurement of power gain of a low-current transistor. The connections use coaxial cables

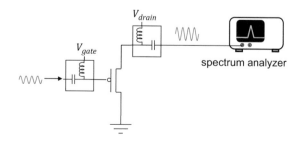

---

[1] The case of gain when the load impedance is high is presented in Chap. 4.

$$A = \frac{P_{\text{out}}}{P_{\text{in}}} \tag{3.1}$$

and can be related to DC measurable quantities by

$$A = \left(g_{\text{m}} \frac{r_0 R_L}{\left(r_0 + R_L \right)} \right)^2 \tag{3.2}$$

where A is the gain, P_{in} is the input power, P_{out} is the output power, g_{m} is the measured DC transconductance, r_0 is the differential output resistance, equal to $1/g_{\text{ds}}$, the measured DC output conductance, and $R_L = Z_0 = 50\ \Omega$.[2] The highest output power is obtained if r_0 is infinite, the ideal case, so $P_{\text{out}} = P_{\text{in}} (g_{\text{m}}R_L)^2$. Very small transistors often have very large r_0. To obtain a power gain of unity or more requires a transconductance of at least 20 mS. When gain is less than one, as in the example presented here, input power must be sufficient to bring P_{out} into the measurement range of the spectrum analyzer.

An example of using this technique to evaluate a top-gated carbon nanotube FET was presented in [1]. Because it was recognized that pad-to-pad capacitance, connected through the substrate, could lead to crosstalk between input and output, the device to be tested was made on quartz, an insulating substrate. As the gate-to-ground capacitance was reduced this way, and microwave probes were used, the input signal amplitude could also be assumed to be fairly constant over a large frequency range, unattenuated by the input admittance.

A measurement of the carbon nanotube FET output power as a function of its input frequency is shown in Fig. 3.6. This shows the power detected at the spectrum analyzer when the gate signal is ±0.3 V, e.g., a moderately large signal (this time, needed in order to achieve an easily measured output, not for the purpose of rectification). The operating biases were chosen to maximize the output swing, based on measurement of DC characteristics. The figure shows that the total measured FET output appears to be constant at low frequency and then starts to increase at about 10 MHz. In addition to measuring the transistor output, measurements were made of the power transmitted between gate and drain by repeating the measurements with the FET in the off-state. It can be seen that this signal, called crosstalk, increases in proportion to the frequency, as expected for a simple capacitance between the input and output. The increase of the FET signal starting at 10 MHz is understood as the measured power being the sum of the FET output and the crosstalk signal. As the contribution of the crosstalk increases at higher frequency, it starts to become much larger than the transistor signal and eventually completely dominates. While the crosstalk signal is independent of the bias applied to the transistor, the FET signal will increase or decrease according to the applied biases, that is, according to the

---

[2] In addition, if the *input* signal line is not terminated in Z_0, then the voltage is doubled at the gate and the gain, which is referred to the power of the signal sent from the signal generator, is increased by a factor of 4, or 6 dB.

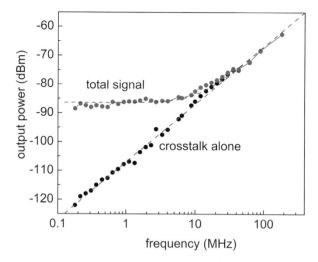

Fig. 3.6 Measured direct power gain *vs* frequency of a carbon nanotube FET [1]. The power applied to the gate is −5 dBm

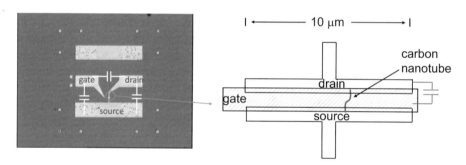

Fig. 3.7 Illustration of the origin of gate-to-drain capacitance of a carbon nanotube FET

DC transconductance, g_m. The amplitude of the input signal can be adjusted to be in the large- or small-signal regime, but the crosstalk signal will increase or decrease in proportion to the input amplitude, so there will be no improvement in signal-to-crosstalk ratio as a result.

In this particular measurement, the source of the capacitance which leads to crosstalk comes from the basic device fabrication, as illustrated in Fig. 3.7, which shows a photograph of the pad layout and a drawing of the transistor contact. Note that the top ground pad of the pads is not used to contact the transistor and is only fabricated to accommodate GSG probes and to provide continuity of the coaxial grounds. As mentioned above, the capacitance between the probe pads was virtually eliminated by the use of a quartz substrate. However, there is a small capacitance between the gate and drain electrodes due to their necessary overlap which is required for fabrication of the carbon nanotube FET, as shown in Fig. 3.7. Although the capacitance due to this overlap is small, about 15 fF, it is much larger than the calculated gate-to-nanotube capacitance of 200 aF for a single nanotube. To take this measurement to higher frequency where crosstalk doesn't dominate clearly requires a change in structure: more carbon nanotubes in parallel on the same electrodes, or a much smaller gate electrode with the nanotube placed controllably and precisely. The capacitance between the gate and drain probes themselves will also contribute to crosstalk, but in when the probes are far apart, as in Figs. 2.15 and 3.7, this is a very minor concern.

However, it might be thought that it is possible to extract some sort of frequency response by subtracting the crosstalk signal from the total signal, after converting the measured power to voltage. If this is done, it is found that the carbon nanotube FET signal is diminished somewhat at the top end of the frequency range. This is due to the effect of phase: there is a phase difference between the capacitively coupled crosstalk and the drain current of the FET. Being a scalar instrument, the spectrum analyzer measures just the total power, and phase information is lost. Subtraction based on total power does not account for the difference in phase of the two signals, which leads to apparent reduction of FET signal when the out-of-phase crosstalk signal becomes large. Figure 3.8 illustrates this subtraction with a calculation using parameters similar to the actual FET. In this illustration, the FET is assumed to have a frequency-*in*dependent current, as shown. The scalar power resulting from this current and the crosstalk signal is seen to look like the data in Fig. 3.6. However, extracting the transistor signal by scalar subtraction of power leads to the *apparent* transistor signal being reduced at high frequency, contrary to the assumed input. Thus, this direct measurement method may lead to a pessimistic conclusion if the crosstalk power is subtracted from the total (transistor + crosstalk) power. However, once the crosstalk capacitance has been determined by fitting the

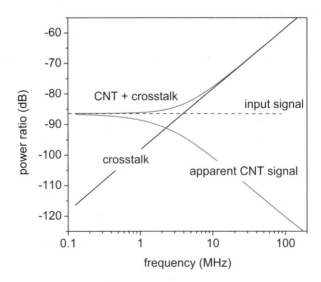

Fig. 3.8 Illustration of the effect of a scalar subtraction of a crosstalk signal from the total measured signal in a power gain measurement

slope of the measured crosstalk signal, it is possible to fit the measured total power to test assumptions about the transistor frequency response. The dashed lines in Fig. 3.6 demonstrate this process. The lines are based on the measured crosstalk capacitance and the assumption of a frequency-independent device output. The match to the transistor data is quite good, suggesting that the assumption that CNFET output is constant over the measured range is valid. However, the sensitivity of the fit clearly diminishes when the crosstalk dominates, and this result can only be considered to show that the measurements of Fig. 3.6 are not inconsistent with a flat FET response over the measurement interval. Though a phase-aware subtraction cannot be done, this direct measurement is also consistent with the high-frequency performance of carbon nanotube FETs demonstrated in the rectification measurement described previously, that the frequency response is unchanged over the measurement range.

To be clear what this measurement shows, it provides a relatively simple means to determine the frequency range of operation, or the minimum value of that range, and is an excellent starting point for a novel transistor. If the transistor signal can be made larger, and the crosstalk smaller, than in this carbon nanotube example, the measurement range will be higher. The measurement does not, however, lead to a recognized figure-of-merit or to an equivalent circuit model of the transistor.

The need to measure phase to correctly subtract parasitic capacitance points to the need for *vector* power measurement. This is the role of the *S*-parameter measurements of frequency response, using a vector network analyzer, which will be discussed in the next section.

3.3 Frequency Performance Figure-of-Merit and Linear Model

3.3.1 AC Linear Model of Transistors

Beyond the demonstration that the transistor has the potential to work at high frequency, it is important to get an accurate, mathematical, description of its frequency dependence. Such a description usually begins with a comparison with a model which has worked well to describe both silicon bipolar junction transistors (BJTs) and MOSFETs. It is known as a pi model, or hybrid-pi model, and is shown in simplified form in Fig. 3.9 [5]. Although most new transistors will likely be of the MOSFET structure, for completeness, both BJTs and MOSFETs are described here. While the model works at DC and AC for BJTs, it is only useful at AC for MOSFETs. It is used here to provide a context for the measurements which are desired.

Note that the hybrid-pi model is a linear network, unlike a SPICE model, so it is useful for small-signal calculations only. It consists of a transconductance component, which is a current source controlled by an input voltage, so that the output current, i_2, is proportional to an input voltage. The input of the devices is a capacitor and resistor in parallel, in the BJT case, Fig. 3.9(a), where C_π is the sum of the

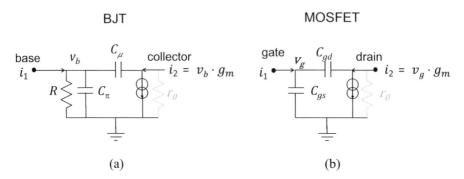

Fig. 3.9 Small-signal hybrid-pi model of transistors. (**a**) Bipolar junction transistor (BJT) and (**b**) metal-oxide semiconductor field-effect transistor (MOSFET)

base-emitter junction capacitance and the emitter diffusion capacitance, C_μ is the base-collector junction capacitance, and the resistance is the differential resistance of the base-emitter junction. In the MOSFET case, Fig. 3.9(b), the capacitance is the gate capacitance, which is the sum of the gate-to-source capacitance and gate-to-drain capacitance, and includes overlap capacitances. Additional fringing capacitances may add to the modeled capacitors. The emitter is grounded for the BJT; the source for the MOSFET. The feed-forward current contribution to the drain current due to C_{gd} or C_μ is relatively small and ignored here. There is also an output resistance, r_0, but in the application of these models to define cutoff frequency, it is assumed that the output is shorted, so this resistance can be temporarily ignored. In the case of FETs, this model is best used for the strong inversion region where the gate voltage applied to the transistor is greater than its threshold voltage and the drain voltage is suitably high. Operation in weak inversion, below threshold, requires a different expression for transconductance, and considering that the transistor performance decreases rapidly as gate voltage is equal to or less than threshold voltage, this regime is not particularly important for assessing device speed.

Accordingly, non-zero DC biases are applied to each terminal, so that even though the output, for example, the drain, must be a small-signal short to ground, the physical drain terminal is *not* grounded. This, of course, is hard to achieve experimentally. It might be thought that a bias tee could provide an AC short while applying a DC bias, but because a bias tee is part of a transmission line, shorting it is not the same as shorting the drain locally, and will have varying impedance due to the quarter-wave and half-wave effects as discussed in Chap. 2. Additionally, a bias tee doesn't operate at low frequency, or DC, in any case. This problem is handled using *S*-parameter transformations, as will be discussed below.

It is implicit in the concept of small-signal models that the transistor is in a stable DC state. With novel devices, that might not be obvious, as application of voltage, or operation in atmosphere, can change device characteristics. Chapter 7 discusses stability measurement techniques in some detail, but for the purposes of small-signal measurements, it may be sufficient to monitor the bias currents during the measurement. This doesn't guarantee short-term stability, but will detect long-term drift.

A figure-of-merit obtained from this model is the small-signal current gain, known as both β and h_{21}, and equal to i_2/i_1, which are the small-signal output and input currents. The reduction of this gain as frequency increases leads to the performance figure-of-merit known as cutoff frequency. In the case of the BJT, which is a current controlled device even at DC, current gain is a physical meaningful quantity, whereas for a MOSFET, it is derived from a physical voltage input, and is only meaningful for AC input. For the BJT, the DC current gain, known as β_0 or h_{fe}, is given by $g_m R$. To calculate AC current gain, the two capacitances, which are both in parallel to ground, are lumped into one term, C.

From

$$h_{21} = \frac{i_2}{i_1} = g_m \left(R \parallel X_C \right) \tag{3.3}$$

it is easy to find that

$$\left| h_{21} \right| = \frac{h_{fe}}{\sqrt{1 + \left(\dfrac{f}{f_\beta} \right)^2}} \tag{3.4}$$

where $f_\beta = \dfrac{1}{2\pi RC}$ is the corner frequency, at which the value of $|h_{21}|$ is reduced by 3 dB from its zero frequency, i.e., DC, value. (The corner frequency occurs at the positive value of a complex pole at $j\omega = -1/RC$, where $\omega = 2\pi f$ is the angular frequency.) The corner frequency of an RC filter, confusingly, is called cutoff frequency, but conventional use of the term in BJT performance has a different definition.

By defining $f_T = h_{fe}f_\beta$, this can be rewritten as

$$\left| h_{21} \right| = \frac{f_T}{f\sqrt{1 + \left(\dfrac{f_\beta}{f} \right)^2}} \tag{3.5}$$

f_T is commonly called the cutoff frequency, though "transit frequency" or "transition frequency" are better terms, and refer to its physical meaning. Cutoff frequency of a BJT is a measure of its frequency response *and* its DC current gain.

By letting R go to infinity, and avoiding the use of h_{fe}, which is undefined (i.e., it is always infinite) for a MOSFET, the current gain of the MOSFET can be determined. It is also easy to derive it directly from the model of Fig. 3.9(b), which has a pole at $\omega = 0$, with the result that

$$\left| h_{21} \right| = \frac{f_T}{f} \tag{3.6}$$

Hence, cutoff frequency is the frequency at which the gain (the absolute value of small-signal current gain with output shorted) is unity. For both the BJT and the MOSFET, it can easily be seen that

$$f_T = \frac{g_m}{2\pi C} \tag{3.7}$$

where the capacitance is the corresponding sum, as described for the BJT above. The capacitance used in the MOSFET version of Eq. (3.7) is the sum of C_{gs} and C_{gd}, commonly known as C_{gate} Hence, the hybrid-pi model predicts how the AC current gain falls as frequency increases. Unlike the BJT, cutoff frequency of a MOSFET does not depend on a DC gain. Furthermore, for both MOSFETs and BJTs, the model establishes a connection between a high-frequency description, f_T, and DC and low frequency measured characteristics, g_m and C. In conventional silicon FETs, C has a fairly constant value over the bias conditions for which the channel is strongly inverted, but for novel transistors this may not be the case. The result expressed in Eq. (3.7) combines two desired properties of a transistor into a single figure-of-merit: gain and high-frequency response. As frequency response depends on the inverse of capacitance, and gain depends on transconductance, cutoff frequency combines these two properties, and hence, is a very popular single number for characterizing performance.

The description above defines cutoff frequency operationally, but it can be shown to have a physical meaning. Qualitatively, the transit time of signal carriers (electrons or holes) across the active region of a transistor must be inversely related to frequency response. When the frequency is lower than the unity current gain frequency, f_T, the carriers travel completely across the active region of the transistor; at higher frequencies, their direction is reversed before the transit is complete, and the gain is less than one. It can be shown mathematically that the reciprocal of the angular cutoff frequency is equal to a transit time. For BJTs, the transit time is that of minority carriers crossing the base and the emitter-base and collector-base junctions; for MOSFETs, it is the time for the majority carriers to cross the channel. Specifically, for MOSFETs, $1/f_T$ is proportional to L/v_d, where v_d, is the drift velocity, and L is the channel length. Drift velocity is equal to mobility times electric field, μE. One of the interesting properties of an electronic material is its mobility, which is often first measured in its pristine or bulk state, and an important question is does the material retain this mobility when it has been processed to make a transistor. Because of its relation to mobility, measured cutoff frequency is a good way to answer this question.

It is important to note that the short-circuit current gain, as predicted here, is not, in general, physical realizable at high frequency, and must be derived from other measurements. However, it is interesting to examine the predicted behavior. The absolute value of the current gain, $|h_{21}|$, is shown in Fig. 3.10 for both types of transistors. The BJT starts out at a DC value and starts falling near the corner frequency, after which it develops a $1/f$ dependence. The MOSFET can be thought of as having a corner frequency of zero, so it has a $1/f$ dependence at all frequencies. As the

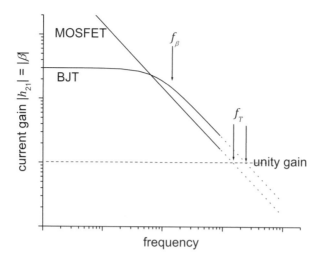

Fig. 3.10 Calculated frequency dependence of small-signal current gain of transistors with output shorted, using the hybrid-pi model

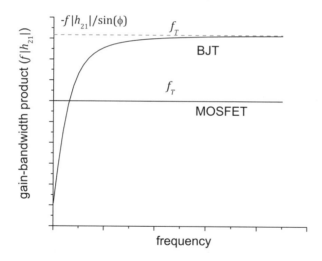

Fig. 3.11 Calculated current gain-bandwidth product of small-signal current gain of transistors with output shorted, using the hybrid-pi model, as a function of frequency

current gain drops, the formulas above predict that it will reach a point of unity gain, which is the cutoff frequency, f_T. Generally, real transistors tend to deviate from this ideal behavior as the applied frequency nears or surpasses cutoff frequency, for a variety of reasons, including the fact that the input to output capacitance has been neglected in the model. And it is often quite true for very fast transistors that cutoff frequency is larger than the maximum frequency of the measurement equipment, so this deviation may not be measured.

It is commonly said that cutoff frequency is the extrapolation of the gain to the frequency at which it has unity value. While qualitatively correct, it is better to think of it as the gain-bandwidth (small-signal, short-circuit current gain) product over the measurement interval. This can be seen in Fig. 3.11, where the absolute value of gain is multiplied by the frequency. For MOSFETs, this results in a straight horizontal line which is equal to cutoff frequency, and which can be read off the vertical axis of a graph. For the BJT, this value is reached asymptotically, unless the magnitude of the current gain is modified by using the phase of the signal, as indicated in the figure.

The hybrid-pi model, applied to the strong inversion regime of an FET, describes three essential attributes of a transconductance-controlled transistor.

1. Its frequency response, as measured by the absolute value of the short-circuit current gain, falls as $1/f$. This provides an operational definition of cutoff frequency.
2. Cutoff frequency depends on the DC transconductance and the net input capacitance, which may depend on the biases applied. As transconductance and capacitance change with DC bias, so does cutoff frequency. (In silicon MOSFETs, capacitance doesn't change much when operated in the saturation regime, so cutoff frequency is predominantly proportional to transconductance.)
3. Cutoff frequency increases in proportion to $1/L^2$, where L is the channel length. This follows from the fact that transconductance increases as channel length decreases, and the geometric fact that gate capacitance decreases with channel length. The length dependence will be weaker if the transconductance is not strictly proportional to $1/L$ (as it isn't for short channel silicon MOSFETs), but f_T will still increase as the channel length is reduced. (This $1/L^2$ relationship also leads to the interpretation that cutoff frequency is inversely proportional to the time for the majority carriers to cross the channel. As mentioned above, that time is L/v_d, where v_d, is the drift velocity, which is proportional V_{ds}/L, where V_{ds} is the voltage across the channel. Hence, channel transit time has a net L^2 dependence.)

If these attributes can be obtained by measurement of the novel transistor, then it is a good indication that the transistor behaves very similarly to conventional silicon field-effect transistors, and means that it is fair to compare their performance. Specifically, cutoff frequency is often cited as a performance figure-of-merit (though it has limited meaning) but this is only a meaningful comparison if the transistor has the features described above. Furthermore, if the new transistor can be defined by a hybrid-pi model, or something similar, then it gives insight into the characteristics which determine its performance, which gives guidance into how to further improve that performance. More broadly, the ability to describe frequency dependence with an equivalent circuit model is an important first step toward building a model which can be used for predictive circuit performance.

3.3.2 Measuring Small-Signal Current Gain with S-Parameters

If the short-circuit, high-frequency current gain cannot be physically realized, how then, can such characteristics be obtained? For this, it is necessary to work with S-parameters. While measurement using S-parameters to characterize high-frequency transistors uses several well-known conventional techniques, it is helpful to collect all of the information in one place.

If the transistor has enough output current, measurements can be made using a vector network analyzer, which directly measures S-parameters. In order to relate S-parameters to transistor parameters, it is necessary to think about the transistor in abstract terms. Imagine that the transistor is a box. It has an input and an output and a common terminal; as each of them is consider a "port," then it is a two-port device [6]. As such, it can be described as having four signals: two input currents and two input voltages. (Currents can be negative or positive, so they can be considered input or output.) This can be pictured as in Fig. 3.12. The box in the figure can be represented by parameters which relate the terminal voltages and currents to each other. These are most easily portrayed as a matrix. For example, a z-parameter representation is:

$$\begin{pmatrix} v_1 \\ v_2 \end{pmatrix} = \begin{pmatrix} z_{11} & z_{12} \\ z_{21} & z_{22} \end{pmatrix} \begin{pmatrix} i_1 \\ i_2 \end{pmatrix} \tag{3.8}$$

This is an impedance matrix representation: each element of the matrix is an impedance, that is, a voltage divided by a current. The impedance matrix describes what terminal voltages will be obtained in response to the terminal currents. The elements of the matrix can be determined by isolating one or the other inputs and can be related to the properties of the two-port device. For example,

$$z_{11} = \left(\frac{v_1}{i_1} \right) \Bigg|_{i_2 = 0} \tag{3.9}$$

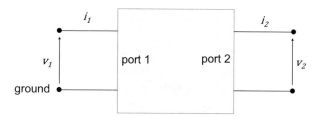

Fig. 3.12 General representation of a two-port network described by voltages and currents

is the impedance of port 1 when port 2 is open circuited (no current can flow). To obtain z_{11}, current is forced into port 1 and its voltage is measured, with port 2 open. Similarly,

$$z_{21} = \left(\frac{v_2}{i_1}\right)\Bigg|_{i_2=0} \qquad (3.10)$$

is the forward transfer impedance (the impedance between port 1 and port 2) when port 2 is open. To make this definition concrete, consider the example of a resistive two-port network as shown in Fig. 3.13. Using the definitions above, it is easy to see that $z_{11} = R_A + R_B$, and $z_{21} = R_B$.

Other parameters can be used to describe the input to output transformation. For example, the y-, or admittance matrix, describes what currents are obtained in response to voltage inputs:

$$\begin{pmatrix} i_1 \\ i_2 \end{pmatrix} = \begin{pmatrix} y_{11} & y_{12} \\ y_{21} & y_{22} \end{pmatrix} \begin{pmatrix} v_1 \\ v_2 \end{pmatrix}. \qquad (3.11)$$

There are six possible sets of parameters to describe a two-port device. In addition to admittance and impedance parameters, the hybrid parameters and ABCD (or transmission) parameters are the most useful. The parameter h_{21} used for current gain in the description of the hybrid-pi model above is one element of the hybrid matrix. An important fact about the parameter sets is that since any matrix describes the two-port completely, any representation can be transformed into any other representation by a linear operation.

Although the example of Fig. 3.13 uses resistors and DC currents and voltages, the description of the two-port device has no such limits. The currents and voltages can be AC, and the elements in the box can be reactive, and the parameters which describe them, complex.

However, physically applying and measuring the input and output currents at high frequency is difficult or impossible. As high-frequency signals travel on transmission lines which have voltages and currents flowing simultaneously, the idea of separately forcing one and measuring the other makes no sense. The seemingly

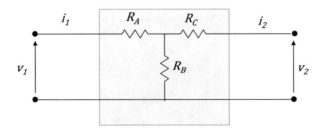

Fig. 3.13 A two-port network of resistors, used as an example to calculate z-parameters

Fig. 3.14   General representation of a two-port network described by S-parameters

static signals shown in Fig. 3.12 cannot be achieved at high frequency. Instead, S-parameters can be used.

S-parameters, or scattering parameters, make use of the idea that AC signals travel on a transmission line of fixed impedance as voltage waves (with current determined by the impedance, Z_0), and that some portion of the wave may be reflected back (scattered) when it arrives at the device being measured. Although S-parameters can describe a two-port network, they do so with incident and reflected *waves* referenced to the device ground, rather than separate voltages and currents, as shown in Fig. 3.14 [6, 7]. Similar to the other matrix representations of the two-port device, the S-parameters form a matrix:

$$\begin{pmatrix} b_1 \\ b_2 \end{pmatrix} = \begin{pmatrix} S_{11} & S_{12} \\ S_{21} & S_{22} \end{pmatrix} \begin{pmatrix} a_1 \\ a_2 \end{pmatrix} \tag{3.12}$$

and

$$a_1 = E_{i1} / \sqrt{Z_0} \tag{3.13}$$

$$a_2 = E_{i2} / \sqrt{Z_0}$$

$$b_1 = E_{r1} / \sqrt{Z_0}$$

$$b_2 = E_{r2} / \sqrt{Z_0}$$

where E is an incident (i) or reflected (r) voltage wave and b_1 and b_2 are normalized output voltage waves and a_1 and a_2 are normalized input voltage waves. The normalization is such that, when squared, these waves represent power in a Z_0 system. The meaning is similar to the other two-port matrices. For example, if $a_2 = 0$, that is, there is no input to port 2, the resulting $b_1 = S_{11}a_1$ can be read as *"the output wave from port 1 equals the input to port 1 times S_{11}."*. That is, S_{11} is a reflection coefficient. All of the S-parameters are ratios of voltage waves.

It is beyond the scope of this book to delve further into the mathematics of S-parameters, but there are two points which make them useful for characterizing devices. First, they describe exactly what a vector network analyzer does: it measures traveling voltage waves impinging on and reflecting from a test device, from either port. Second, they can be linearly transformed into any of the other two-port parameter sets explained above. Except for all the messy details of the measurement, which follow below, the means of measuring h_{21} to test the hybrid-pi model predictions, is simply a matter of measuring S-parameters with a VNA, and converting the S-parameters to hybrid parameters, including h_{21}, which is used in the hybrid-pi model. The VNA provides stimulus and measurement of small-signals applied to the transistor. As in the direct power measurement, the stimulus and response are delivered and measured using microwave probes, while DC voltages are applied from source and measurement units, or simple power supplies. Unlike the apparatus used for direct power measurements described earlier in this chapter, bias tees are usually built in to the VNA.

3.3.3 Calibration

However, measuring with a VNA requires calibrating with known standards and must be done often. Calibrating a VNA is not the same as a factory calibration of a voltmeter or oscilloscope, something done every few years by a trained expert. Rather, it is a procedure done to account for phase change, imperfections and losses due to the cables, connectors, and probes which are used to connect the VNA to the devices being tested. A better term for this process is vector error correction, but the term calibration is used universally. This calibration must be done every time the instrument is used after a long idle period, such as days or weeks, or if a cable or probe is changed. Calibration consists of measuring the response to known standards for which the ideal response is known. In the case of RF probes, which are used for the work described here, the probes contact known standards, typically shorts, 50 Ω resistors ("loads"), opens, and throughs (port 1 to port 2 connections). Probes used for S-parameter measurements must *not* have the 50 Ω termination mentioned as a possibility in Chap. 2. In principle, any three known standards can be used, but the short, load, and open are useful as they span the full phase and magnitude ranges of the VNA, so there is no extrapolation. Qualitatively, the short determines the end of the transmission line, the load measures its attenuation, and the open measures fringing capacitance. At each contact, the VNA measurement is stored, and when all are complete, an error model is generated on the instrument and can be applied to all subsequent measurements. The result of that is that the probe tips become a measurement reference plane, and measurements at that reference plane are made as if the VNA and its cables and connections constitute a perfect instrument. In other words, all of the phase shifts and losses of the cables and probes have been calibrated away. If the cables or probes are changed, or the measurement frequencies are changed, a new calibration must be performed.

The use of microwave probes makes calibration fairly easy, as it only requires making probe contact with an insulating substrate, such as sapphire, on which the short, load and through, and sometimes, open,[3] standards have been fabricated.

---

[3] Usually, the open is defined as the probes separated from the wafer, i.e., not making contact with anything.

Such standard substrates are sold commercially, typically by probe manufacturers. The calibration measurements can be done using the VNA front panel buttons, or using commercial external software [8] on a controller which downloads the error models which correct raw data, so the user sees only the "perfect" measurement. It is good practice to take the precaution of measuring the DC value of the standard as it is being used, as the calibration depends on the physical standard matching the value assumed by the VNA. A worn standard, or poor probe contact with the pads can result in mismatched resistance and faulty calibration. For example, if the 50 Ω load standard is actually, say, 55 Ω at DC, then the polar plots, shown below, will be off-center, and all "calibrated" measurements will be wrong. The resistance of the standards is easily measured through the bias tee when the probes make contact with the calibration substrate, but the resistance of the bias tees, typically 1 or 2 Ω, must be subtracted to get the net calibration standard resistance.

An example of some results after calibration is shown in the polar plot of Fig. 3.15. After calibration, measurements of the calibration standards should show almost perfect results. Figure 3.15 shows the reflection coefficient when probing the short, 50 Ω load, and open. The dots shown represent the measurement over the entire swept frequency range of the measurement. The open shows a reflection of unity amplitude, with no phase shift. The short shows a reflection of unity amplitude with a phase shift of 180°, and the load shows zero reflection. (Note that some methods of defining the open result in a small negative capacitance due to probe to substrate capacitance. This is dealt with by the calibration substrate manufacturer.)

In order to be operating in a linear range, it is best to keep the input power fairly low, such as about −25 dBm. This power level should be maintained for the

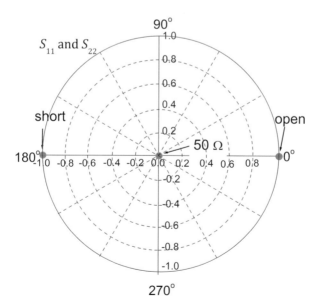

Fig. 3.15 Ideal measurements of short, open, and matched load of a VNA after calibration

calibration and de-embedding steps. Different power levels can be tried, if necessary, to check that the response is the same. As mentioned previously, *S*-parameters are all ratios, so the same result will be obtained at other power levels as long at the device response is linear. This is easily tested experimentally.

As mentioned at the beginning of the chapter, if the transistor has a very small output signal, measurement noise may make it impossible or impractical to measure with a VNA, which was why the direct power measurement technique was used to evaluate the carbon nanotube FET described earlier. Estimating the noise level of a VNA measurement is a bit tricky. While the *S*-parameters are ratios, which are unchanged as input levels are changed, the signals actually detected by the VNA do depend on the input strength. For a transistor with a gain in 50 Ω which is substantially less than unity, and a small-signal level for input, the detected signal, specifically the forward gain, S_{21}, can be near the noise floor. Making matters worse, the transistor gain decreases with frequency, leading to greater noise as cutoff is approached. Furthermore, at high frequency, losses in cables may reduce signal amplitude, which on average is recovered by calibration, but adds noise. And the de-embedding process which will be described below may result in the subtraction of two noisy signals from each other, resulting in an even noiser final result. Reducing the IF bandwidth of the VNA, and using its signal averaging capability can reduce the noise, but at a cost of very long measurement time. A certain amount of experimentation may be required to determine if a small transistor can be measured with low enough noise to compare it to traditional FET response.

Since the voltages are applied to the transistor through the bias tees of the VNA, it is possible, and wise, to monitor the currents during the *S*-parameter measurements, to assure that good contact is maintained and that the transistor is stable. It is also often a good idea to measure I-V curves, through the bias tees, before starting a measurement. It must be noted, though, that some VNAs have resistive shunts to ground, typically 1 MΩ, on their input/output ports. These shunts add extra current to the current measured, which should be subtracted to get correct device currents. This subtraction can lead to current measurement errors when the transistor being tested is very small.

3.3.4 De-Embedding Procedure

While calibration is necessary for any VNA measurement, it is not sufficient for measuring transistor response. It is also necessary to account for the probe pad environment surrounding the transistor in order to obtain the behavior of the transistor alone. This can be done through modeling of the probe pads, but is much better treated by the process of de-embedding. It is important to keep the concepts of calibration and de-embedding separate. Calibration results in an apparently perfect instrument, but de-embedding is still required to extract the transistor's behavior from the environment in which it is fabricated and tested.

Fig. 3.16 Drawing of transistor enclosed by probe pads. (**a**) The small square at the center represents the transistor and (**b**) a model of the effect of the probe pads to surround the transistor with capacitors and resistors

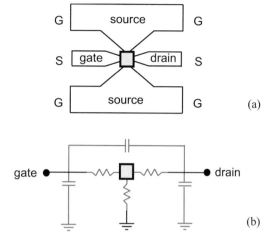

(a)

(b)

A successful calibration brings the reference plane to the tips of the microwave probes. At this point, they contact the probe pads which are connected to the device. These probe pads, however, add unwanted parasitic components to the measurement. As illustrated in Fig. 3.16, the transistor device is a microscopic feature inside macroscopic pads, so it is embedded in the capacitance and resistance of the probe pads and connections. The capacitances are between the pads, predominantly connected by the underlying substrate, though, as discussed in the carbon nanotube FET example, some unwanted gate-to-drain capacitance can be built into the transistor structure. Some resistance can come from the pads themselves, but it is dominated by the very narrow metal lines which connect the pads and the transistor. In a good design, the parasitic resistance is very small, but the pad capacitance is unavoidable. Note that the illustration is a simplified model of the most important parasitic connections.

A measurement of such a structure measures the response of the transistor embedded in the parasitic elements, which may be considerably different from the response of the intrinsic transistor. The goal of the process called de-embedding is to make additional measurements, on special probe structures, which make it possible to empirically remove the capacitances and resistances from the total measurements, revealing the intrinsic transistor behavior. This process, particularly the subtraction of capacitance, is possible because of the vector nature of the measurements which include phase.

The full de-embedding process requires that two additional structures be fabricated along with the devices. These are the "open" and "short" structures shown in Fig. 3.17. Both use the same probe pad layout as the transistor, except that the transistor is absent, and the open has unconnected pads where the transistor would be, while the short has all of the pads connected at the same point. When measuring with the network analyzer, in addition to measuring the S-parameters of the device, those of the open and short are also measured, and used in the subsequent analysis. As suggested by Fig. 3.17(a), use of the open data removes the capacitance, but the

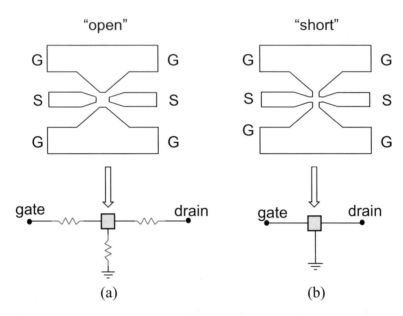

Fig. 3.17 Design of open and short de-embedding structure to match the total test structure shown in Fig. 3.16

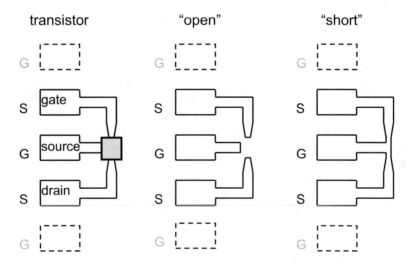

Fig. 3.18 A compact probe pad layout for use with SGS or GSGSG microwave probes

connection resistances remain, and in (b) use of the short data removes the resistances, *if* the open correction has been done first.

The transistor, open, and short designs shown in Figs. 3.16 and 3.17 are widely used in the research and commercial communities, but they do consume a lot of space on the wafer. When space is either not plentiful, or very expensive, there is a more compact layout which can be used. This is illustrated in Fig. 3.18. Instead of using two opposing GSG (ground-signal-ground) probes, this design uses a single SGS, or GSGSG, probe in a linear array. For modest frequencies, say, below about 10 GHz, the SGS probe works well. At higher frequencies, it becomes necessary to shield the signal probes on both sides, so the GSGSG layout is preferred. Because of its linear arrangement, adjacent devices can share the outer G pad, to conserve space.

The de-embedding process uses straightforward linear algebra (though with complex numbers) with the two-port matrices, but because there are many steps, it appears complicated. The mathematical steps are listed in Fig. 3.19. The parasitic capacitances add admittances to the transistor response, and the parasitic resistances add impedances between the pads and the transistor, so they are removed by subtraction using the admittance matrix, or y-parameters, and the impedance matrix, or z-parameters, respectively. The process makes heavy use of the transformation between the various two-port matrices described above. The algebraic operations refer to the matrices.

1. The "raw" measured S-parameters, which are those of the transistor surrounded by pads, S_{raw}, are converted to y-parameters, y_{raw}.
2. The measured S-parameters of the "open" structure are converted to y-parameters, y_{open}.
3. The open y-parameters, y_{open}, are subtracted from the raw measured transistor parameters, y_{raw}.
4. The result of the subtraction is converted to S-parameters, S_{de1}, where the nomenclature designates that the first de-embedding has been done.

Fig. 3.19 Step-by-step procedure to perform the open and short device de-embedding using the structures of Figs. 3.17 and 3.18

$$1. \quad S_{raw} \rightarrow y_{raw}$$
$$2. \quad S_{open} \rightarrow y_{open}$$
$$3. \quad y_{de1} = y_{raw} - y_{open}$$
$$4. \quad y_{de1} \rightarrow S_{de1}$$

} "open" subtraction

$$5. \quad S_{short} \rightarrow y_{short}$$
$$6. \quad y_{shde} = y_{short} - y_{open}$$
$$7. \quad y_{shde} \rightarrow z_{shde}$$
$$8. \quad S_{de1} \rightarrow z_{de1}$$
$$9. \quad z_{de2} = z_{de1} - z_{shde}$$
$$10. \quad z_{de2} \rightarrow S_{de2}$$

} "short" subtraction

At this point, the S_{de1} parameters represent the transistor response *de-embedded* from the parasitic capacitance [9]. It is now possible to operate on these parameters to extract the required information, such as cutoff frequency, input and output impedance, gate resistance, and maximum frequency of oscillation. In many cases, this is adequate de-embedding, and the next steps to be performed do not change the response much. However, in the case where the connection resistance is high, or the current is high leading to iR drops, the z-subtraction should be performed:

5. The measured S-parameters of the "short" are transformed to y-parameters, y_{short}.
6. Similar to the subtraction from the transistor above, the capacitances are subtracted from the short y-parameters, resulting in a short de-embedded from the capacitances, y_{shde}.
7. These latter are converted to the z-matrix representation, z_{shde}.
8. The de-embedded device parameters from step 3 above are converted to z-parameters, z_{de1}.
9. The capacitance de-embedded "short" z-parameters, from step 7, are subtracted from the capacitance de-embedded transistor z-parameters.
10. The z-parameters from step 9 are transformed to S-parameters, S_{de2}, where the nomenclature now indicates that two levels of de-embedding have been done.

With that, the S-parameters of the transistor are de-embedded from parasitic capacitance and resistance of the probing structure. At this point, further figures-of-merit can be derived from the de-embedded S-parameters. Specifically, cutoff frequency is obtained by the final step of transforming the de-embedded S-parameters to hybrid parameters, and using the h_{21} element of the matrix, which is the input to output current gain. In the hybrid parameter representation, h_{21} has the output short circuited, as required for the hybrid-pi model.

3.3.5 Measurement Examples

An example of a typical measurement is shown in Fig. 3.20, which shows the magnitude of the current gain, $|h_{21}|$, of a silicon FET. The solid circles show the gain obtained from de-embedding. In Fig. 3.20(a), the logarithm of gain is plotted directly against the logarithm of frequency and can be seen to fall with $1/f$, as expected from the hybrid-pi model. The solid line is an extrapolation to the unity-gain frequency, f_T. Figure 3.20(b) shows the same gain data multiplied by frequency, $f(|h_{21}|)$, as suggested in Fig. 3.11, resulting in a constant value. It is easy to read the value of cutoff frequency on the vertical axis. This measurement example also shows how cutoff frequency can be thought of as gain-bandwidth over the measurement range. As cautioned above, measurement noise of a low-current device is not likely to yield a result as quiet as this silicon FET example.

The effect of the de-embedding can be seen in these data and is quite pronounced. Speaking roughly, the magnitude of de-embedding is governed by the size of pad

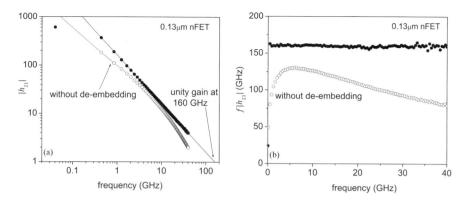

Fig. 3.20 Measured cutoff frequency data for conventional silicon nFET. (**a**) Plot of current gain, h_{21}, *vs* frequency for the raw data (open circles) and the data treated with the de-embedded procedures (solid circles) and (**b**) the same data plotted in the gain-bandwidth form

capacitance compared intrinsic capacitance, such as the gate capacitance. If the pad capacitance is relatively large, then a large admittance is added to the incoming signal, and the signal arriving at the intrinsic device is reduced, changing its apparent frequency response. The open circles show the data before de-embedding the probe pad capacitance. Clearly only the de-embedded data follow the expected $1/f$ form. The gain-bandwidth plot in Fig. 3.20(b) emphasizes how large is the effect of pad capacitance. The shape difference seen depends upon the relative magnitudes of the transistor capacitances and of the parasitic capacitances and cannot be generalized to follow a particular form. Though de-embedding compensates for the problem without resorting to any model of the parasitic effects, a large amount of de-embedding is undesirable as it will increase measurement noise. The best practice is to make the gate and drain pads as small as possible, particularly if the transistors are very small. An insulating substrate is also desirable, if it can be used.

Using the calibration and de-embedding methods, the goal of this work can be tested: to see if the novel transistor behaves like a conventional MOSFET or BJT. An example is drawn from the early work with graphene FETs [10]. Graphene was the first two-dimensional material to capture public attention, in part because of its extraordinarily high mobility in its pristine state. Although many DC studies of graphene FETs can be made with the back-gated structure, it is fairly easy to make a graphene FET with a top gate, using the structure in Fig. 3.1(b), where the channel material is a deposited flake of graphene.

The DC characteristics of such an FET are quite different from its silicon counterpart. As one example, the output characteristics (I_d–V_d curves) of a silicon nFET and a graphene FET are compared in Fig. 3.21. However, graphene transistors may have much more output current than carbon nanotube FETs (mA *vs* μA) so they deliver much more AC signal amplitude, making VNA measurements possible.

Measurements were made of the S-parameters of the graphene transistor, and the manipulations described above, which de-embed the transistor and convert S-parameters to obtain h_{21}, were performed. An example of the resulting current gain, $|h_{21}|$ is shown in Fig. 3.22. It can be seen that the first requirement to validate

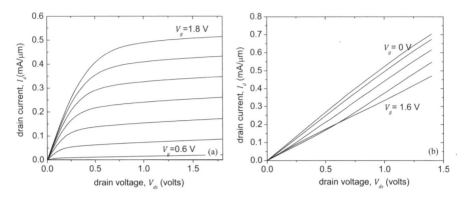

Fig. 3.21 Measured output characteristics ($I_d - V_d$ curves) of FETs. (**a**) Conventional silicon nFET and (**b**) exploratory FET made with a graphene channel

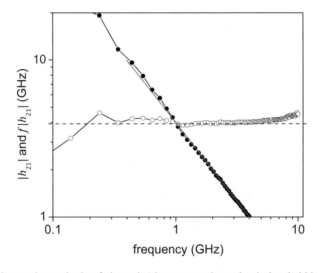

Fig. 3.22 Measured magnitude of short-circuit current gain, and gain-bandwidth product, of a graphene FET

the hybrid-pi model is satisfied: the gain falls with frequency, as $1/f$. Figure 3.22 also displays the gain-bandwidth product, $(f| h_{21}|)$ as discussed in connection with Fig. 3.20. It can be seen that $f| h_{21}|$ is not perfectly flat, but rises a bit at the higher frequencies. It is noted that some of the frequencies at which the gain is measured are higher than the cutoff frequency, 4 GHz in this example, which explains the excess signal. Referring to the hybrid-pi model, Fig. 3.9, there is a capacitance between the gain and drain which was ignored in defining cutoff frequency. However, when the transistor gain is less than unity above cutoff frequency, and the transmission through the capacitance simultaneously increases with frequency, that capacitance can contribute significant signal to the output, causing an increase

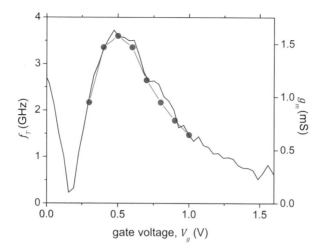

Fig. 3.23 Measured cutoff frequency and transconductance of a graphene FET, with arbitrary scaling factor for the overlay, as a function of gate voltage [9]

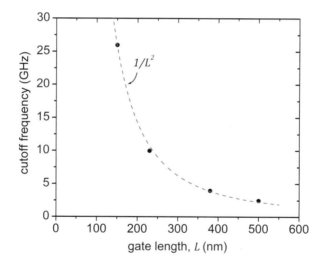

Fig. 3.24 Measured cutoff frequency of a graphene FET as a function of gate length [9]. The maximum cutoff frequency at each gate length is used for the plot

above the expectation of Eq. (3.6). This example shows that it is generally wise to measure only to frequencies which are less than the expected cutoff frequency.

The second test of the hybrid-pi model is the correlation of f_T and g_m. An example of this test is shown in Fig. 3.23, which superimposes curves of each quantity plotted as a function of applied gate voltage. It is seen that the two track each other quite well, which makes the case that, in spite of very different DC behavior, and a very different channel conduction mechanism, a graphene FET has a small-signal frequency behavior which is similar to silicon FETs.

Finally, the dependence on gate length is shown in Fig. 3.24. The maximum f_T measured with devices of different channel length is shown, and, as seen by the imposed line, it increases as the gate length is reduced, in proportion to $1/L^2$, as anticipated from the model. These three results show that the small-signal behavior of the graphene FET behaves very similarly to a silicon FET, in spite of the very different channel physics.

3.3.6   Definition of the Physical Open Structure

In Fig. 3.17, the open is illustrated as a structure which looks like the transistor in its pad set (Fig. 3.16) but with the transistor snipped out. This can be called a "pad open," and is easy to define and commonly used. In some cases, frequently encountered with new channel materials, it is also possible to make an open by simply removing, or, actually, not placing, the channel material. Thus, in Fig. 3.1(b), the entire transistor structure would be built, except for the channel material. This is electrically an open, in that there is no active device, and any signals measured passing through this structure are due to inter-pad capacitance and whatever internal capacitances remain when the channel material is not present. This can be called an "intrinsic open." This method is simple and has been used extensively in graphene measurements.

However, while both kinds of opens serve the purpose of subtracting parasitic capacitance in the de-embedding technique above, they account for different amounts of capacitance subtraction, and thus, lead to different device results. Although the y-subtraction method described in Fig. 3.19 is a purely "black box" subtraction which makes no assumptions about the structure of the transistor, one effect of the subtraction can be seen, as in Fig. 3.16, as reducing the net gate capacitance. A smaller gate capacitance results in higher cutoff frequency, as in Eq. (3.7). The pad open subtracts capacitance due only to the probe pads. If the substrate is insulating, or of high resistance and the oxide layer is thick, so that gate-substrate capacitance is small, the overall gate capacitance is dominated by the gate-to-drain and gate-to-source overlap area and using the intrinsic open for de-embedding also subtracts these additional components. As no realistic transistor can be built without these internal capacitances, use of the intrinsic open is sometimes regarded as an over-subtraction, resulting in an unfair measure of performance.

The significance of the open definition in measuring performance is shown in an example in Fig. 3.25, which displays the gain vs frequency plots of a graphene FET. The data shown are from the same FET; the different curves result from the type of open which is used in the de-embedding. This seemingly small difference creates a performance figure-of-merit, f_T, which differs by more than a factor of two. The intrinsic open naturally results in higher cutoff frequency.

It is also possible to create a pseudo-open by biasing the transistor to turn it completely off. This is a convenient way to do an open subtraction when a separate open structure has not been designed and fabricated. In the off-state, the signal flows only through capacitances, which may be predominantly the probe pads, but also include

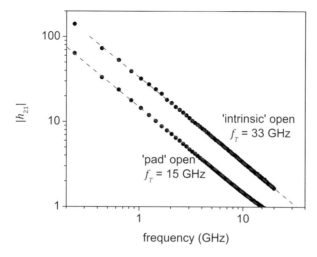

Fig. 3.25 Measured gain vs frequency of a graphene FET using two different physical "opens" for de-embedding

internal capacitances, so it is somewhat like the intrinsic open as described above. However, the capacitance of the off-device includes intrinsic gate capacitance, which should not be removed by de-embedding. If the substrate is not insulating, as in the usual silicon transistor FETs, a substantial gate capacitance remains even when the channel is depleted. In this case, de-embedding using the off-device clearly subtracts too much internal capacitance. Off-state de-embedding can be useful for a preliminary measurement when a proper test structure is unavailable, but will always leads to some amount of de-embedding error and should generally be avoided. However, use of the off-state was used to obtain the purely capacitive signal in the carbon nanotube FET direct power measurement shown in Fig. 3.6 above, when proper vector parasitic subtraction could not be applied and the purpose was only to compare the effect of a purely capacitive signal transmission to that of an active transistor. And also, in that example, the substrate was insulating.

Which is correct, intrinsic open de-embedding or pad open de-embedding? This is a matter of intent and interpretation. Generally, the pad open is preferred. But as explained above, cutoff frequency is related to channel mobility. The intrinsic performance measures how well only the channel material works, without regard to contact structures, and leads to evaluation of the material transport. For example, does its cutoff frequency correspond to the mobility of the material before it is processed? It also establishes an upper limit for cutoff frequency. The pad open performance, in contrast, measures the performance of the entire fabricated device, which include elements like capacitance, and perhaps resistance, which reduce performance, but which might be subject to change by layout or fabrication method, and therefore may not be optimal.

The important point is that the physical definition of the open actually defines what part of the test structure is the device. This is not a matter of right or wrong de-embedding, but a definition of where the device begins and the test connections end. For some diagnostic tests, this distinction may not be insignificant, but in reporting frequency performance, it is important to be clear.

3.3.7   Other Performance Figures-of-Merit

From the set of S-parameters, it is possible to determine other frequency and gain figures-of-merit which are of interest to the circuit design community. A commonly cited figure-of-merit is "maximum frequency of oscillation," f_{max}, which is the power analog of cutoff frequency, the frequency at which a power gain falls to unity. Power gain falls, however, as $1/f^2$. Several kinds of power gain can be defined, such as simple power gain, available gain, maximum available gain, maximum stable gain, unilateral gain, transducer power gain, and unilateral transducer power gain. The various power gains are defined by the assumptions about matching the input and output impedances to achieve maximum power transfer and may not be physically realizable. Even more than cutoff frequency, power gains depend heavily on transistor layout, and considerable design and simulation effort is required to maximize power gain from a given intrinsic transistor. As this is relevant to amplifier and circuit design and is not a study of transistor fundamentals, is beyond the scope of this book.

References

1. D. V. Singh, K. A. Jenkins, J. Appenzeller, D. Neumayer, A. Grill and H.-P. Wong, "Frequency response of top-gated carbon nanotube field-effect transistors," *IEEE Transactions on Nanotechnology*, vol. 3, no. 3, pp. 383-387, Sept. 2004, DOI: https://doi.org/10.1109/TNANO.2004.828577.
2. P. J. Burke, "An RF circuit model for carbon nanotubes," *IEEE Transactions on Nanotechnology*, vol. 2, no. 1, pp. 55-58, March 2003, DOI: https://doi.org/10.1109/TNANO.2003.808503.
3. S.-J. Han, K.A. Jenkins, A. Valdes-Garcia, A.D. Franklin, A.A. Bol, and W. Haensch, "High-Frequency Graphene Voltage Amplifier," Nanoletters, vol 11, no. 9, pp. 3690-3693, Aug. 2011, https://doi.org/10.1021/nl2016637.
4. D. J. Frank and J. Appenzeller, "High-frequency response in carbon nanotubefield-effect transistors," *IEEE Electron Device Letters*, vol. 25, no. 1, pp. 34-36, Jan. 2004, DOI: https://doi.org/10.1109/LED.2003.821589.
5. Y. Taur and T.H. Ning, *Fundamentals of Modern VLSI Devices*, Cambridge: Cambridge University Press, 1998.
6. T.H. Lee, *The Design of CMOS Radio-Frequency Integrated Circuits*, Cambridge: Cambridge University Press, 1998.
7. D. M. Pozar, *Microwave Engineering, 2nd Edition*, New York: John Wiley & Sons, 1998
8. "WinCal" product of FormFactor, Inc., Livermore, CA; www.formfactor.com
9. M.C.A.M. Koolen, J.A.M. Geelen and M.P. J.G. Versleijen, "An Improved Deembedding Technique for On-Wafer High-Frequency Characterization," IEEE 1991 Bipolar Circuits and Technology Meeting, pp. 188–191 (1991).
10. Y.-M. Lin, K. A. Jenkins, A. Valdes-Garcia, J.P. Small, D. B. Farmer, and P. Avouris, "Operation of Graphene Transistors at Gigahertz Frequencies," *Nano Letters*, vol 9, no.1 , pp. 422-426, Jan, 2009, DOI: https://doi.org/10.1021/NL803316h.

Chapter 4
Case Studies in the Evaluation of Novel Transistors

This chapter takes up two problems which are related to high-frequency performance, but do not fit neatly into the categories of Chap. 3. The first is the establishment that a transistor can have useful AC voltage gain, over some frequency range, when the new transistor is too small to be expected to drive a low impedance load. In this case, it might still be useful as a voltage amplifier in a hybrid technology. Measurement of voltage gain, however requires a special technique. The second problem is related to the effect of gate resistance on the power gain frequency response. In a conventional technology which is modified to improve the gate resistance, anomalous resistance can occur, which hurts the performance, but which is undetectable by DC measurements. This resistance can only be detected by using a VNA as a diagnostic instrument.

4.1 Voltage Gain of a Low-Current Transistor as a Function of Frequency

Whereas cutoff frequency is a figure-of-merit which describes the fundamental signal propagation delay of the intrinsic transistor and does not assess voltage or power gain at all (since the output is a short- circuit), and maximum frequency of oscillation describes the power gain with specific input and output impedance matching which may not be physically realistic, a high value for either figure-of-merit does not assure that a real transistor will provide high-frequency gain when driving a realistic load. However, there is another situation of interest when the load is very high or infinite, and *voltage* gain is achieved. Not only is this useful for characterization, but such a situation often arises in analog and digital circuits. In a multi-stage amplifier, for example, some of the transistors just drive another single transistor, and as such, the DC load presented to a first transistor is the high-impedance input of the next transistor. This situation is analogous to the pre-amp/

© Springer Nature Switzerland AG 2022
K. A. Jenkins, *RF and Time-domain Techniques for Evaluating Novel Semiconductor Transistors*, https://doi.org/10.1007/978-3-030-77775-3_4

power-amp distinction used in audio electronic circuits, where the pre-amplifier boosts the voltage and the power amplifier drives speakers. Voltage gain is of particular interest for transistors which have very low current, resulting in gain of less than unity in low impedance measurement equipment, but which might be useful in circuits where they can be used to drive high-impedance circuit components. A novel transistor which has voltage gain over a large frequency range might be of interest for use in a hybrid technology or circuit.

Voltage gain as a function of frequency extends the DC concept of intrinsic gain to the frequency- domain. A figure-of-merit which is often applied to transistors for analog use is the intrinsic gain, also known as self-gain, K, which describes the *open-loop* DC small-signal voltage gain, that is, output voltage relative to input voltage when the output is open-circuited, i.e., an infinite load. This is in contrast to cutoff frequency, which is short-circuit current gain, and maximum frequency of oscillation which uses power gain. In FETs, using DC-measurable quantities, intrinsic gain is defined as:

$$K = \frac{dV_d}{dV_g}\bigg|_{open} = \frac{dI_d}{dV_g}\frac{dV_d}{dI_d} = \frac{g_m}{g_{ds}} \tag{4.1}$$

The g_{ds} term is the output conductance, which is the slope of the output $(I_d - V_d)$ characteristic such as in Fig. 3.17, and g_m is the slope of the I_d–V_g curve. In an ideal silicon MOSFET, operating in saturation, g_{ds} has a value of zero, or, if its reciprocal, r_0, is used, it has infinite output resistance. Intrinsic gain is a device parameter which represents the maximum differential DC voltage gain which the device can provide and depends on the bias values. Modern MOSFETs seldom have output conductance even approaching zero, and exploratory devices are also unlikely to exhibit a saturated output characteristic, so this figure-of-merit is used to compare the amplifying prospects of different technologies.

Voltage gain derives from the voltage which the intrinsic output resistance, r_0, develops as the current, which is caused by transconductance, is generated when the output is open. Typically, a transistor will not be considered useful if its intrinsic gain is less than about 5. Note that intrinsic gain is defined and measured with the output, the drain, open-circuited. If a resistive load is applied to the drain, it can be shown that the achievable voltage gain, A_V, is given by the self-gain times a factor which depends on the load and on the output conductance, and which is less than unity for finite loads. This can easily be obtained from Fig. 4.1, which uses the same hybrid-pi model used in Chap. 3, but now allows for a finite load, R_L, on the output, instead of the shorted output used to derive cutoff frequency. Ignoring capacitance, and recalling that $r_o = 1/g_{ds}$, the DC voltage gain is

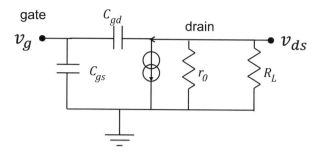

Fig. 4.1 Hybrid-pi model of FET for use when output is not shorted to ground

$$A_{\mathrm{V}} = \frac{\mathrm{d}V_{\mathrm{d}}}{\mathrm{d}V_{\mathrm{g}}}\bigg|_{\mathrm{loaded}} = -\frac{g_{\mathrm{m}}}{g_{\mathrm{ds}}}\left(\frac{R_{\mathrm{L}}}{R_{\mathrm{L}}+\dfrac{1}{g_{\mathrm{ds}}}}\right) = -K\left(\frac{R_{\mathrm{L}}}{R_{\mathrm{L}}+r_0}\right) \qquad (4.2)$$

This is essentially as presented in the discussion of power gain in a 50 Ω environment in Chap. 3, but recast to show that intrinsic gain, K, is the maximum voltage gain of an FET, since any applied load causes a reduction of A_{V}. If the load resistance is infinite, an open circuit, the maximum gain is obtained. If the load resistance is zero, the voltage gain is zero, but note that this is short-circuit condition used for cutoff frequency, where current gain is measured and the output voltage is zero. Intrinsic gain is independent of the FET width, as the same drain current is in both the numerator and denominator. The term in parenthesis, however, does depend on device width, and for a finite load, such as 50 Ω, gain depends on load resistance and g_{ds}. Thus, K is a more fundamental figure-of-merit from which other gains can be obtained. A small FET is likely to have a large output resistance and for this case, the output conductance term can be ignored, and the gain simplifies to:

$$A_V = -g_{\mathrm{m}}R_{\mathrm{L}} \qquad (4.3)$$

Note that the gain described above is derived for DC conditions with a resistive load. While intrinsic gain is easily obtained from DC measurements, its frequency dependence is also needed. In principle, this gain can be computed from measured S-parameters, converting to y-parameters and using the expression

$$K(\omega) = -\frac{y_{21}}{y_{22}} \qquad (4.4)$$

which gives the intrinsic gain for any network for which these y-parameters are known. However, in practice, this method only works for well-established FETs which have enough current to lead to low noise S-parameter measurements. As applied to the hybrid-pi model of an FET, Eq. (4.4) can be related to DC parameters and internal device capacitance. In this case, open-circuit gain as a function of frequency becomes

$$K(\omega) = -\frac{g_m}{g_{ds} + j\omega C_{gd}} \tag{4.5}$$

where C_{gd} is the gate-to-drain capacitance, and $\omega = 2\pi f$, the angular frequency. This expression is valid only if C_{gd} is small. From this, it is easily seen that the magnitude of the gain falls as:

$$|K(\omega)| = \frac{g_m}{g_{ds}} \frac{1}{\sqrt{1 + \dfrac{\omega^2 C_{gd}^2}{g_{ds}^2}}} \tag{4.6}$$

Similar to the mathematical form of current gain of the bipolar junction transistor, Eq. (3.5) and Fig. 3.10, which is that of a one-pole low-pass filter, the self-gain starts at a DC value, $K(0)$, and above a corner frequency, falls as the reciprocal of frequency.[1] While this expression might also appear to be a useful method to measure AC self-gain, it is not easy to separately measure gate-to-drain capacitance, as distinguished from total gate capacitance, particularly when C_{gd} is very small in comparison, even for well-established silicon MOSFETs, so the expression is not very useful in predicting the performance, and a direct measurement is needed.

Hence, measuring voltage gain of experimental transistors requires new techniques. By definition of the open-circuit voltage gain, the output cannot be attached to a spectrum analyzer or power sensor since they present 50 Ω loads. For sufficiently low frequencies, a high impedance, 1 MΩ input, oscilloscope can be used to measure the output while the input is stimulated with a swept-frequency sinusoid. However, such oscilloscopes typically have high capacitance inputs, which limit their bandwidth. In addition, they may not be sensitive to small-signals.

An alternate method which has been demonstrated is to use high-impedance active probes [1]. Such probes are usually used to diagnose integrated circuits. The probe is a very fine wire and can be described as a whisker, which is attached directly to a high gain and bandwidth amplifier with a high input impedance. Because of the high impedance of the amplifier, the whisker does not present a resistive load to the device or circuit and measures voltage. The amplifier has an *effective* gain of 0.1, but its purpose is not to amplify, but to provide a signal large enough to drive 50 Ω while not loading the device being probed. There is some small capacitive load, which is specified by the manufacturer. The probe is positioned with a high-quality manipulator. Such a whisker can be placed on a small metal line on a

---

[1]The same mathematical form occurs when there is an external load, which explains why the power gain of the carbon nanotube FET driving a spectrum analyzer, shown in Chap. 3, appears flat over the modest frequency range of the measurement even though current gain may fall continuously over the entire frequency range.

Fig. 4.2 Photograph of microwave probes used for voltage gain measurements. The transistor is contacted with microwave probes in the usual way, and a high-impedance active probe, drawn in grey, is placed on the input or output probe pads

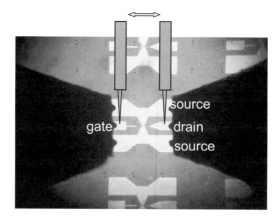

digital circuit, and the amplifier output is connected to a 50 Ω instrument, so the signal on the line can be measured. Although usually used for the purpose of measuring digital logic signals in integrated circuits for diagnostic purposes, a high bandwidth version (26 GHz) of the probe is available which can be used for analog measurements, too.

If the amplifier is connected to a spectrum analyzer, then the whisker can be placed on the input gate metal pad and output drain pad, as shown in Fig. 4.2. Either two such probes can be used, or one can be moved from gate to drain as needed.

An example of the technique to determine the voltage gain of graphene FET as a function of frequency has been demonstrated [2]. The particular FET studied was made using the buried gate technology described in Fig. 3.2. It used an especially thin gate oxide, which resulted in greater degree of saturation of the output current than previous graphene FETs. As most graphene FETs show very little gain because of their high output conductance, e.g., lack of saturation, as seen in Fig. 3.21(b), measurement of this better device was of particular interest. The complete measurement situation is illustrated in Fig. 4.3. The transistor was probed with the microwave probes shown in Fig. 4.2, and operating biases were applied with bias tees. The gate was driven by a signal generator, and the input and output signals were measured with the active high-impedance probe and spectrum analyzer. Gain is simply the ratio of the measured output and input signal amplitudes obtained with the high-impedance probe.

It is to be noted, though, that contacting the transistor with a microwave probe to which a bias tee is attached unavoidably connects the drain to a section of unterminated, that is, open, transmission line. As shown in Chap. 2, an unterminated transmission line presents a frequency- or length-dependent impedance at its input. The transmission line cannot be terminated, as that would create a 50 Ω load, which defeats the purpose of a voltage gain measurement. However, this frequency-dependent transmission line impedance can be used to advantage. When an ideal transmission line is unterminated, that is, open at its end, recall that if the length of the line is l, and λ is the wavelength of the sine wave, the impedance of the input of the transmission line equals its termination for the half-wave line, that is, when

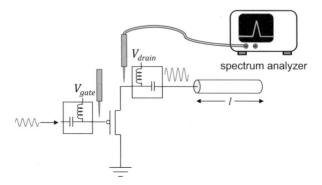

Fig. 4.3 Illustration of the apparatus used to measure voltage gain *vs* frequency. All of the connections use coaxial cable, and an additional short length of cable is attached to the bias tee to cause the impedance of the drain connection to alternate between high and low impedance, as discussed in the text

$$l = n\frac{\lambda}{2} \tag{4.7}$$

and n is an integer. The impedance of an open line thus oscillates between infinite and lower values as the wavelength changes. High impedance occurs when the frequency is a multiple of $v/(2l)$, where v is the speed of propagation in the transmission line. A small section of transmission line in the form of coaxial cable is attached to the output of the bias tee. Its length can be chosen to provide the required density of measurement points for the total desired frequency range. For example, a total length of 50 cm of typical coaxial cable, where the propagation speed is 0.7 times the speed of light, or 2.1×10^{10} cm/s, the high-impedance peaks repeat with a period of 210 MHz. A longer line, which allows longer wavelengths, results in smaller frequency increments between the high-impedance peaks. Too short a length of line, which might occur if no extra cable is added to the bias tee, might prevent the observation of any of the high-impedance peaks.

The transistor is DC-biased to operate at approximately the maximum transconductance. It is assumed that the transistor is stable when operated for moderately long times at fixed DC bias. Voltage gain is measured by the application of a small-signal sine wave, but to obtain the largest output signal, for best measurement accuracy, the input amplitude is maximized. By observing the gain at a single frequency, the optimal input strength can be chosen. This is shown in Fig. 4.4, which shows the output signal at a single frequency as a function of input signal amplitude. Over a very large range of more than four orders of magnitude, the output power is proportional to the input, but eventually starts to deviate at higher power, as seen in Fig. 4.4(a). This is shown more distinctly by plotting the gain in Fig. 4.4(b). This departure from proportionality is called gain compression and is one means of measuring device linearity, which is discussed in Chap. 5. Using inputs below the gain compression point ensures small-signal operation. The input signal level was set at

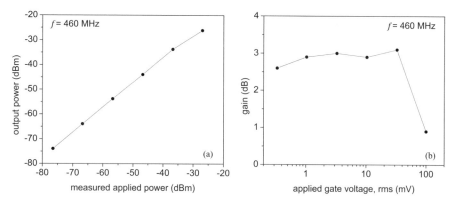

Fig. 4.4 Measurement of gain to establish optimum operating point for the voltage gain of a graphene FET. (**a**) Output power *vs* input power and (**b**) gain *vs* input power [2]

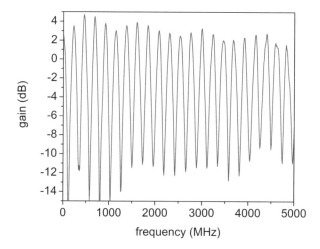

Fig. 4.5 Gain of a graphene FET as a function of frequency, measured with the apparatus shown in Fig. 4.3 [2]

30 mV rms (−17 dBm) for the full frequency-dependent measurement of this graphene FET.

The gain is measured as the input signal frequency is swept with small enough steps to resolve the peaks, with the result shown in Fig. 4.5. The gain clearly oscillates between high and low as a periodic function of frequency, as expected from the impedance variation of an open-terminated transmission line as the frequency passes repeatedly through the quarter wave and half-wave conditions. Only the peaks of the gain are the desired voltage gain, since for frequencies in between the peaks, the input impedance of the transmission line is not infinite, and the FET drives a finite (and complex) load. In addition to the gradual reduction of gain at

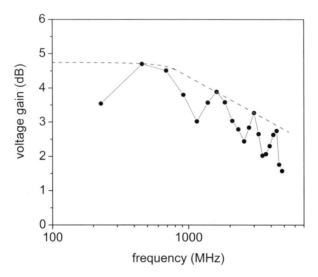

Fig. 4.6 Gain of a graphene FET under high-impedance load conditions. These are the data from only the peaks of the data of Fig. 4.5. The dashed line is a prediction of gain based on S-parameter measurements

higher frequencies, there is some additional structure, which is probably due to reflections from imperfect components.

Figure 4.6 shows the *peak* gains *vs* frequency, extracted from Fig. 4.5. The additional structure is apparent, but it is noted that the true voltage gain may be greater than any point in the Figure because voltage gain is defined for infinite impedance, whereas the impedance of a *real* half-wave transmission line, one with losses, can be finite. The dashed line in Fig. 4.6 is drawn to show the frequency shape expected from Eq. (4.6), above, for $|K(\omega)|$, starting at a DC value and then starting to fall with frequency.

Since the impedance of a half-wave line is infinite only for a perfect transmission line, lossy lines can be expected to show lower impedance. This is shown explicitly in Fig. 4.7, where, in a separate measurement, a network analyzer is used to measure the reflection coefficient of the input end of the same transmission line used for the voltage measurement. From the measured S_{11}, the impedance is calculated using

$$Z = Z_0 \left(\frac{1 + S_{11}}{1 - S_{11}} \right) \tag{4.8}$$

This magnitude of this impedance is plotted in Fig. 4.7, and shows that although the impedance has maxima at periodic frequencies, as predicted from half-wave theory, some of the peak values are only a few times the characteristic impedance of 50 Ω. As remarked above, this means that the true voltage gain must be greater than the gain measured in high, but not infinite, impedance. Knowing the impedance from the data shown in Fig. 4.7 make it possible to correct the measured data in

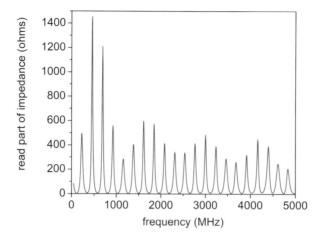

Fig. 4.7 Measured real part of the impedance of the output transmission line in the voltage gain apparatus, as a function of frequency

simulation, and in the experiment reported, the simulated gain is slightly higher than that of Fig. 4.6, and smooth, and indicated that voltage gain greater than unity persisted, for this experimental graphene device, to 15 GHz. There is no equivalent of de-embedding for a measurement of this type, so the gain measured and simulated includes the effects of probe pad parasitic capacitance. For comparison, the raw, un-de-embedded, cutoff frequency for this transistor was about 8 GHz.

In spite of the limitations seen by using coaxial cables which are not ideal, the work shows that intentionally adding a transmission line to load the output makes it possible to obtain the open circuit voltage gain as a function of frequency, which is helpful for evaluating the utility of a novel small transistor. Without the extra added transmission line, the measured, apparent, gain would have been lower, and wrong. By exploiting the half-wave nature of transmission lines, it is possible to measure voltage gain as a function of frequency.

4.2   Vertical Gate Resistance

This section shows the use of frequency analysis as a tool, not to measure performance of a transistor directly, but to diagnose the nature of a component of the FET which can affect its performance. Specifically, the gate resistance is analyzed by frequency measurement.

The resistance of the gate electrode of an FET is an important parameter in determining the speed or frequency of the transistor. In the hybrid-pi model used to define cutoff frequency, the input gate is modeled as a pure capacitor which is a good assumption for modeling a forced AC current used in the definition. In the real physical situation, though, there is unavoidable resistance in the gate, between the external terminal and the gate-to-channel capacitor. When the gate is made of poly-silicon, its resistance may not be negligible. In the case of digital CMOS circuits, this gate resistance can slow down the propagation delay between stages. Stage-to-stage delay can be seen as due to the rather simple concept of RC delay. A first stage drives the capacitance, the gate, of the next stage. The driving stage drives this capacitance with its drain current, but the current can be seen as the reciprocal of the output resistance of the driving stage, so the net delay is due to the output resistance of the driving stage times the input capacitance of the following stage. Of course, because of changing voltages, regarding the output resistance and input capacitance as constants is vastly oversimplified, yet it provides a good guide for scaling and sizing of logic circuits. However, this simple picture neglects the resistance of the gate electrode of the driven stage. If that resistance is significant compared to the output resistance of the driving stage, then the stage-to-stage delay is increased.

In the case of AC operation of a transistor, the gate resistance has an effect of the maximum frequency of oscillation. In a simplified model, the maximum frequency is given by

$$f_{max} = \sqrt{\frac{f_T}{2\pi R_g C_{gd}}} \tag{4.9}$$

where R_g is the gate resistance and C_{gd} is the gate-to-drain capacitance. While f_T is a measure of the intrinsic device, f_{max} includes all the resistances which affect extrinsic, total performance when input and output are impedance-matched. Gate resistance is often considered extrinsic to the transistor because it doesn't directly control channel conduction and can potentially be modified without changing the active region of the transistor. High gate resistance clearly lowers the performance of a single FET. Considerable effort is spent to maximize the maximum frequency by optimizing device layout to minimize the denominator in this expression.

Large gate resistance also acts to increase the noise figure of transistors and circuits.

The scaling which has occurred over the last decades, which has been driven by the relentless desire to build faster transistors, as well as to increase their packing density, has resulted in an undesirable effect on gate resistance: as to the channels become smaller, resulting in higher drain current (lower output resistance), the gate electrode becomes narrower, and its resistance necessarily increases, potentially undoing or even reversing some of the gains achieved by the channel length reduction. The increased gate resistance caused by dimensional scaling can slow down digital logic gates and reduce f_{max}, in spite of the higher drain current and cutoff frequency.

In order to prevent such problems from arising as further scaling continues, there has been considerable effort to engineer lower gate resistance, in addition to optimizing layout. Most effort has concentrated on replacing some or all of the conventional gate material, polysilicon, which is fairly resistive, with a lower resistivity metals, usually with some structural changes such as layering. Significantly, these efforts are evaluated with DC measurements of the resistance of test structures consisting of long gates with two or four terminals, for Kelvin measurements. The result of this measurement is the quantity known as sheet resistance, or sheet-rho.

However, in a layered or composite gate electrode, this measurement does not elucidate the existence of a vertical resistance, which is the resistance from the top of the electrode to the gate dielectric. While in a uniform polysilicon gate, this resistance can be estimated from the traditional measurement, in a layered structured it might be completely unknown.

Consider the structure illustrated in Fig. 4.8. The figure shows a hypothetical gate electrode composed of two layers. Between the layers is an interface layer, which may be resistive or even insulating. The usual measurement of sheet resistance, which is from one end of the gate to the other, reveals the total end-to-end resistance, which can be described as the horizontal resistance, R_h. If there is resistive interface, measured horizontal resistance will predominantly due to the resistance of the top layer. The resistance from the top of the gate to the contact to the gate dielectric, the vertical resistance, R_v, is not measured, and *cannot* be measured, by DC measurements which force a current and measure the resulting voltage, as there is no complete DC current path. Yet this resistance can be of paramount importance in determining the net gate resistance which can affect the performance. In an extreme case, if the interface layer is insulating, no gate signal will get to the channel, even if the horizontal resistance is low. Of course, the resistances and capacitance are distributed along the width of the gate, and the representation as discrete elements in Fig. 4.8 is just for illustrative purposes.

However, the gate dielectric provides an AC path through from the top of the gate to the source and drain contacts, and the impedance of the gate dielectric decreases as the AC frequency increases. The dielectric will become electrically transparent at sufficiently high frequency. The net impedance of the path from gate to source or drain can be measured using the reflection coefficient (S_{11}) of a VNA (vector network analyzer), which proves to be a very useful instrument for diagnosing such effects. The VNA measurement can be considered somewhat analogous to the

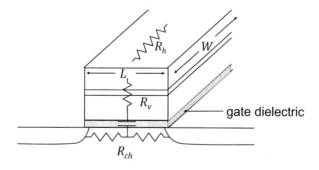

substrate

Fig. 4.8 Illustration of the different components of gate resistance

4-point DC measurement of resistance measurement in which resistance is measured by forcing a small current and measuring the resulting voltage, using Kelvin sensing to eliminate contact resistance. Measurement with a VNA forces a small AC signal and measures the reflected voltage wave which indicates the gate's impedance and de-embedding is used to remove the effect of the connecting probe pads.

4.2.1   Detecting Gate Discontinuity Using VNA Measurements

An example of detecting a vertical gate resistance problem was demonstrated in [3]. An fabrication experiment was done to determine the importance of the gate resistance on propagation delay, as measured with ring oscillator frequency. A composite gate with low resistance was compared with two simple gate structures, as controls, by using the same basic device technology and changing only the gate structure. The wafers were processed in parallel to avoid lot-to-lot variations which could cause differences in ring oscillator frequency. The composite gate had a top thick metal layer ($TiSi_2$), to lower the net resistance, and a thick bottom layer of polysilicon to contact the device channel. To prevent metal diffusing from the top layer to the gate oxide, and thereby changing the threshold voltage, a thin conducting layer of TiN was grown between the layers to act as a diffusion barrier, with the resulting structure looking, in cross-section, conceptually like Fig. 4.8, but with the barrier layer meant to be conducting. The horizontal resistance, the sheet resistance, of the composite gate was more an order of magnitude lower than the control gates, at all gate lengths measured, so it looked like the best gate to achieve fast signal propagation.

However, measurements of ring oscillators showed, surprisingly, that the devices with this lower sheet resistance, which were expected to be faster, actually had significantly greater propagation delay than the controls, and they didn't have the usual dependence on channel length. The cause of this apparent discrepancy was determined by measuring the total gate impedance with a VNA operated in the GHz range. Large area devices were available for easy study with a network analyzer. Unlike the two-port transistor measurements described in Chap. 3, only a single parameter, S_{11}, was measured. The transistor was biased "on" so that the channel was conducting in order to minimize channel resistance, R_{ch}, and one-port open de-embedding was used to remove the effect of pad capacitance. (Figure 3.19, steps 1–4, applied to S_{11} only.)

A typical plot of the resulting reflection coefficient of a control gate as a function of frequency is shown in Fig. 4.9, which indicates that the total impedance to ground is that of a resistor in series with a capacitor. S_{11} is converted to z_{11}, which equals $R + X_C$ (after adjusting the z-parameter for 50 Ω normalization) in the model for Fig. 4.9. The capacitance component of the total impedance is extracted from the frequency dependence of z_{11} these for transistors of various gate lengths. The qualitative result is shown in Fig. 4.10, which shows capacitance/area of the gate as a function of gate length. The control gates show the capacitance per area increasing

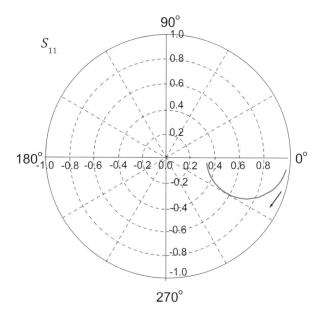

Fig. 4.9 Typical VNA measurement of the reflection coefficient of an FET gate with the source and drain grounded. The frequency range of the measurement is from .045 to 40 GHz

Fig. 4.10 Measured normalized capacitance of the gate electrode of FETs as a function of gate length [3]. Data are shown for a novel gate structure and for a conventional gate used as a control

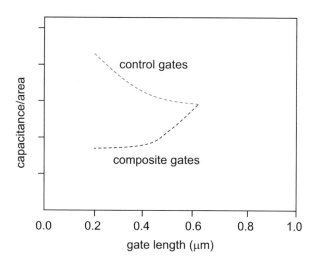

at short gate lengths because edge capacitance adds relatively more capacitance for shorter gates. The edge capacitance is that due to the capacitance between the vertical sides of the gate to the planar source and drain below, which adds to the gate oxide capacitance. Without this extra edge capacitance, capacitance per area would be independent of gate length.

The composite gates, in contrast, show the capacitance per area *decreasing* for shorter gates. This is explained by the proposed equivalent circuit shown in Fig. 4.11. In addition to the horizontal and vertical resistors in series with the gate oxide capacitor, a simple explanation of the composite gate capacitance Fig. 4.10 is the existence of an additional fixed capacitor, C_d, at the location of the TiN barrier layer. An approximately fixed capacitor can explain the capacitance per area of the composite gate. As the gate gets narrower, this capacitor becomes greater relative to the gate oxide capacitance, so the net measured capacitance per area decreases.

This discontinuity capacitance was hypothesized to be due to unwanted formation of TiO on the sides of the barrier layer. Fig. 4.12(a) illustrates a cross-section of the ideal gate structure with TiN forming fully across the gate, but in this conjecture, as illustrated in Fig. 4.12(b), TiO forms at the edges, and penetrates a fixed distance, creating an extra, and constant, capacitance as in the schematic in Fig. 4.11, a capacitance which becomes relatively greater for narrower gates. This hypothesis was reached without any destructive physical analysis, just using vector network measurements. The TiO explanation was subsequently verified with TEM (transmission electron microscope) images of cross-sections of gates.

The TiO penetrated to a length of about 0.2 μm. This almost completely isolated the lower portion of the gate stack for the shortest of the gates. Although analyzed and identified using capacitance as a metric, the measurements lead to the conclusion that the vertical *resistance* of the shortest channel devices is a significant portion of the total gate resistance. This was corroborated by time-domain measurement of signal delay through individual FETs using an oscilloscope. It was seen that the short channel FETs with the composite gates showed large RC charging time constants when compared to the control gate FETs. Although a reduced capacitance might be expected to speed up the transistor, the corresponding effect of increasing the vertical resistance outweighs the effect of reducing the total capacitance.

Fig. 4.11 Equivalent circuit to explain the anomalous dependence of gate capacitance on length. In this model, the excess capacitance C_d and resistance R_v are assumed to be constant, not scaling with gate length

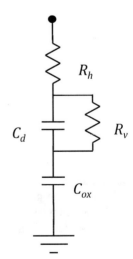

Fig. 4.12 Schematic cross-section of the gate structure to explain the physical origin of the extra capacitance proposed for the equivalent circuit of Fig. 4.11. (**a**) Desired gate structure and (**b**) origin of the excess capacitance due to oxidation of Ti at the sides of the gate

Hence, a frequency-domain technique is able to reveal a partial discontinuity in the composite gate stack, which is fundamentally unmeasurable by conventional DC methods, and it demonstrates the potential importance of vertical resistance in such structures.

4.2.2 Measuring Vertical Gate Resistance in Continuous Gate Stack

Even if there is no change as dramatic as the unintended growth of an insulating region in the gate stack causing a mechanical interruption of resistance, a large, but DC-undetectable, vertical resistance can still occur in a complex composite gate stack. Such a circumstance calls for recognizing a signature in the total resistance alone, which may be much more subtle than a marked change in capacitance seen above.

Such a circumstance arose in the development of a stacked layer gate in a standard CMOS process. The gate structure was as shown in the schematic cross-section of Fig. 4.13 [4]. The gate was a high-κ, metal-gate type, where a polysilicon layer was formed on top of a metal layer. In this structure, there is the potential for increased resistance at the interface between the two layers, as indicated by the arrow. To make a good ohmic contact between the upper and lower materials, the gates were implanted with boron, after formation of the polysilicon layer. As in the first example, above, ring oscillators seemed to run more slowly than expected from the known device characteristics and the measured DC sheet resistance. Fabrication experiments were done with different thicknesses of polysilicon, in order minimize the sheet resistance, and VNA measurements were used to search for excess vertical resistance.

As in the gate discontinuity study above, the VNA applies a small-signal, swept over frequency, and the reflection coefficient is measured. Unlike the structure above, there were no open pads available for de-embedding, so an open was simulated by operating the device in the off-state. As this makes the channel region highly resistive, it acts like an open made as if the conducting channel were removed. Unlike the cutoff frequency measurement, such an open does not lead to misleading estimate of performance. The capacitance extracted from the resulting measurement may be affected, but as the object is to measure the resistance of the gate, the

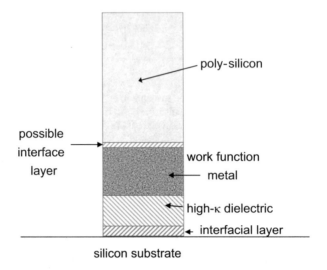

Fig. 4.13 Schematic cross-section of the components of the complex gate stack for use with high-κ gate dielectric [4]. Drawing is not to scale

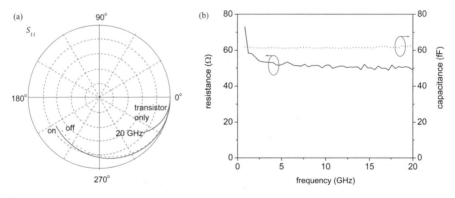

Fig. 4.14 Typical VNA measurement of the reflection coefficient of an FET with the complex gate stack. (**a**) Measured polar plot of FET embedded in the probe pads, biased on (solid line) and off (dotted line). The off-state is used to de-embed the reflection coefficient of the FET from the probe pads. (**b**) The resistance extracted from the real part of the de-embedded reflection coefficient and the capacitance from the imaginary part

off-state de-embedding is an acceptable procedure. A measurement example of the "on" and "off" signals, and the result of de-embedding using the "off" as an open structure, is shown in Fig. 4.14(a), which shows quite dramatically how much phase the probe pads add to the reflection signal. The de-embedded S_{11} has the polar form of a resistor and capacitor in series.

To extract resistance, the de-embedded S-parameter, S_{11}, is transformed, as above, to the z-parameter, z_{11}, which is the total (complex) impedance of port 1 to ground. The resistance is simply the real part of z_{11}, times Z_0, and the imaginary part is the capacitive impedance, $1/(\omega CZ_0)$. The extraction of these parameters from the data of Fig. 4.14(a) of such measurements is shown in Fig. 4.14(b), where the capacitance is calculated from the imaginary part of the impedance. Both the resistance and capacitance are almost independent of frequency in this example, so the simple *RC* model describes the situation well. Although the discrete model works well here, in general, due to the distributed nature of resistance and capacitance which results from the geometry of a wide gate, this model will not fit exactly. Note that this measurement measures the *total* effective gate resistance, not just the vertical component.

The same measurements and analysis were performed to understand the effect of different polysilicon thickness in the experiment described. Data are presented in Fig. 4.15 for the four different polysilicon thicknesses in the experiment. As noted above, if the resistor and capacitor of the gate were single, discrete components, the real part would be completely independent of frequency, that is, flat lines, but because of the distributed nature of the gate, the equivalent series resistance starts to drop at high frequency. Measurement noise accounts for the droop at low frequency: the total impedance, $X_C + R_g$, is very high so a small measurement error can result in a large error in resistance when separating out the real component. Overall, however, this method clearly shows that three of the four data have much higher resistance than the others.

This is a result of the effect of the boron doping: when the polysilicon is too thick, the implanted dopants don't have enough energy to reach the target interface region so they are unable to create a low resistance contact between the two upper and lower layers.

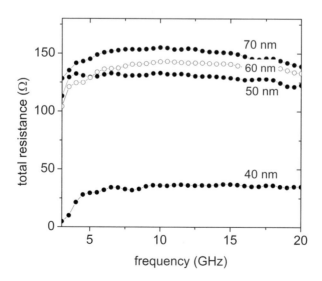

Fig. 4.15 Measured total resistance of the gate stack of Fig. 4.13 as a function of frequency, for different thicknesses of the upper layer of polysilicon [4]

As remarked above, the resistance measured here is the total resistance, which is the sum of horizontal and vertical components. In this example, where a variation in polysilicon height leads to a strong difference in resistance, the addition of a vertical component is easily identified. In general, however, such a separation is not obvious, and requires choosing appropriate transistor dimensions to emphasize the differences [5].

The various resistances in the gate have different dimensional dependences, with vertical resistance being proportional to the inverse of the area of the gate, and horizontal resistance being proportional to its width and inversely proportional to its length:

$$R_h \propto \frac{W}{L} \tag{4.10}$$

$$R_v \propto \frac{1}{LW}$$

$$R_{ch} \propto \frac{L}{W}$$

where L is the gate length (L_g) and W is its width. In units, horizontal resistance depends on the sheet resistance (Ω/square), while vertical resistance depends on resistivity ($\Omega \, \mu m^2$). Neglecting the channel resistance, which can be reduced by applying biases to the transistor, the horizontal and vertical components have opposite dependence on the width, which can lead to one or the other terms dominating.

Figure 4.16(a) illustrates the tradeoff implied by Eq. (4.10). Both horizontal and vertical components are shown as a function of hypothetical gate width, along with their sum. It is seen that for small gate widths, which reduce the contact area, the

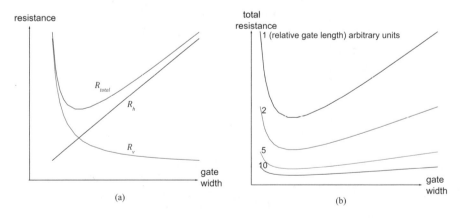

Fig. 4.16 Analysis of the dimensional dependence of the components of gate resistance. (**a**) Horizontal, vertical, and total resistance of gate as a function of gate width and (**b**) Dependence of the total resistance on gate width for various gate lengths. The labels indicate the gate length, in arbitrary units

vertical resistance will dominate; large widths, in contrast, cause the horizontal resistance to dominate, so the total resistance exhibits a minimum as the gate width is varied. Large gate widths should not be used if a vertical resistance is to be measured, and gates should be of a size which puts the total resistance to the left of the minimum in Fig. 4.16(a). If a variety of gate widths are fabricated, it is easy to plot the total resistance as a function of width to see if this minimum is covered in the range of fabricated widths.

Examples of total resistance curves *vs* gate width are shown in Fig. 4.16(b) for different hypothetical gate lengths. Although it may appear that the shorter gates accentuate the difference between horizontal and vertical components, the *relative* division between the two is not affected by gate length. However, the longer gate transistors have lower total resistance, which may make measurement more prone to error. It is clearly advisable to consider the device dimensions to make accurate measurements and to enable easy separation of vertical and horizontal components. While the smallest total gate area results in the largest vertical resistance, it is also important to keep in mind that with a small transistor, de-embedding can introduce errors if two similar numbers (pad capacitance plus device and pad capacitance alone) are subtracted to find a small difference (the de-embedded transistor).

Even in the absence of an expected vertical resistance problem, for example, in a simple gate stack, the AC measurement of gate resistance use the VNA can still be quite useful. The sheet resistance measured with DC structures can be used to predict the resistance of the gates in the fabricated FETs, if dimensional effects and processing differences have no effect. If there is no vertical resistance, the VNA measurement should yield the resistance predicted from the sheet resistance, taking into account the distributed nature of the AC resistance and capacitance of the structure (DC resistance divided by three if contacted from one end and the resistance is less than the capacitive reactance). As such, it is a useful measurement to confirm that sheet-rho is a good predictor of gate resistance, and that there is no vertical resistance of significance.

References

1. Model 35 Picoprobe, product of by GGB Industries, Inc., Naples, FL 34101; www.ggb.com
2. S.-J. Han, K. A. Jenkins, A. Valdes-Garcia, A. D. Franklin, A. A. Bol, W. Haensch (2011) High-frequency graphene voltage amplifier. Nanoletters 11(9), 3690-3693. doi: https://doi.org/10.1021/nl2016637
3. K. A. Jenkins, J. N. Burghartz, P. D. Agnello (1996) Identification of gate electrode discontinuities in submicron CMOS technologies, and effect on circuit performance. IEEE Transactions on Electron Devices 43(5), 759–765. doi: https://doi.org/10.1109/16.491253.
4. K. A. Jenkins, K. Balakrishnan, D. Lee, V. Narayanan (2019) AC Device variability in high-κ metal-Gate CMOS Technology. IEEE Electron Device Letters 40(1), 13–16. doi: https://doi.org/10.1109/LED.2018.2882268.
5. R. A. Wachnik et al. (2014), Gate stack resistance and limits to CMOS logic performance. IEEE Transactions on Circuits and Systems I: Regular Papers 61(8), 2318–2325. doi: https://doi.org/10.1109/TCSI.2014.2321199.

Chapter 5
Measurement of the AC Linearity of Transistors

The subject of linearity, or its complement, non-linearity, of the response of an electronic component to an input signal is well known and documented, but it has usually been considered a subject of concern for analog circuits. However, it is of interest to determine the linearity of the novel transistors which are intended for use in circuits since starting with highly non-linear components makes the design of linear circuits very challenging. In addition to providing performance information and design guidelines, however, the measured non-linearity can contribute to the understanding the fundamental physics of the transistor action.

This chapter describes two conventional methods of measuring non-linearity, but of transistors, not circuits, and specifically, AC non-linearity. The emphasis is on the practical aspects of the measurements, which are often neglected in traditional texts. The connection is made between the measured non-linearity, and the DC characteristics, which are governed by device physics.

5.1 Linearity Overview

It is particularly interesting to understand the linearity of novel transistors, as they may differ significantly from tradition MOSFETs or BJTs. For example, the mechanism of channel conduction of carbon-based transistors, such as carbon nanotube FETs or graphene FETs, is quite different from that of a silicon transistor, being essentially a Schottky-controlled or gated resistor, respectively. As a resistor is a highly linear element, transistors made with carbon might be more linear than conventional semiconductor-based devices. But this must be examined with measurements, so techniques of measuring linearity, though well-known, are presented here in that context.

Linearity was already briefly discussed in an early chapter in distinguishing between small-signal and large-signal operation of a transistor. The regime of small-signal operation is defined by the range of stimulus where the transistor can be

© Springer Nature Switzerland AG 2022
K. A. Jenkins, *RF and Time-domain Techniques for Evaluating Novel Semiconductor Transistors*, https://doi.org/10.1007/978-3-030-77775-3_5

represented by a linear network, that is, a network where the output, whatever it is, is linearly proportional to the input. The concept of a linear representation of a transistor was the basis of the hybrid pi-model which is used extensively in Chaps. 3 and 4. But linearity, or, rather, non-linearity, is not just a matter of measurement and modeling, but has implications for many analog circuits. In audio circuits, for example, non-linearity can lead to distortion of the original sound (although this may be attractive for electric guitars), and in communication circuits, it can lead to false and interfering signals. Consequently, assessing the fundamental linearity of novel transistors is an important endeavor.

Linearity is principally addressed by two measurements: gain compression and third-order intercepts. These are specifically AC concepts and measure the linearity of the response of the transistor to RF inputs. They are also known by the effects they have, which are harmonic distortion and intermodulation distortion (IMD), respectively. Although not always expressed this way, the difference between the two measurements is in the number of distinct frequencies, often called tones, which are applied to the device or circuit: compression is found from a one-tone measurement and third-order intercept is determined from a two-tone measurement.

5.2   One-Tone Non-Linearity: Gain Compression

Gain compression measurement is a simple technique in which a single frequency is applied to the circuit, or, in this case, to the transistor, and its output is measured. The measurement setup is the same as shown in Fig. 3.5, but it is possible to replace the spectrum analyzer with a power meter for this measurement. In this scalar measurement, the strength of the input signal is increased from very small to the point at which the output power is no longer proportional to the input power. The value of the input at which the output deviates from linearity by 1 dB is called the 1 dB compression point [1]. (1 dB corresponds to a 26% change of power, or 12% change of voltage.) The measurement is illustrated by the example of a graphene transistor in Fig. 5.1, which was used in Chap. 4, to set the maximum input voltage to a carbon nanotube FET [2]. The output power is proportional to the input power until it reaches −30 dBm, at which point the gain starts to reduce, that is it becomes compressed. The slope of the linear portion of the curve is, of course, the gain. In this example, the 1 dB point corresponds to an input voltage of 85 mV rms, which is actually quite a large signal. Eventually, if the input power is large enough, the output will stop increasing at all, that is, the device will saturate. (That is, unless the transistor breaks first.) Even if the exploratory device has low or negative gain when loading it with the 50 Ω input of the analyzer, compression can still be measured as the reduction of gain. But as 1 dB is a fairly small change to observe, the measurements must be made with good accuracy. This can usually be achieved by signal averaging. Gain compression is a simple and directly measured quantity.

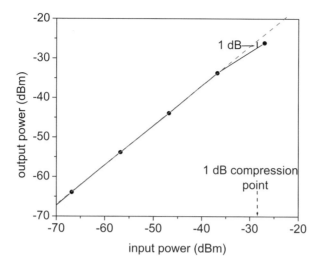

Fig. 5.1 Measurement and description of gain compression figure-of-merit

The distinction is made repeatedly in this book, and elsewhere, between large-signal and small-signal regimes, where small-signal is regarded as the condition for which the response is linear. This gain compression figure-of-merit is a quantitative definition of what regime of operation the device acts in a small-signal manner. Linearity of a transistor depends on its DC characteristics. As required for other measurements, the stability of the DC behavior is assumed in order to measure, and to define, its linearity. In measuring linearity of circuits, a fixed power supply voltage is assumed, but for a transistor, a linearity figure-of-merit also depends on the operating biases, and a single number, though informative, does not present the full description of linearity. This is easy to see why this is the case for the one-tone situation by examining the transfer curves of the device. For example, Fig. 5.2(a) illustrates the drain current *vs* gate voltage of a silicon nFET. The three curves are for three values of the drain voltage, as shown. At 1 V, it looks like the current is linearly proportional to the gate voltage over a fairly large span. It appears that operating in the mid-point of that gate voltage span with a small-signal will generate a linear response, but increasing its amplitude will cause compression because the gate voltage will swing into the subthreshold region or into the roll-off at high gate bias. At 0.6 V on the drain, the same is true, but the apparent linear range is much smaller. At 0.2 V, there is simply no region of the curve which is linear. In fact, a closer examination shows that the transistor is not linear anywhere in these voltage ranges. Figure 5.2(b) shows the transconductance, which is the slope of the transfer curve, for all three voltages. Although it is possible to draw a straight line through much of the 1.0 V curve and declare that it is linear, the graph of transconductance proves otherwise. Of course, the question of linearity is a question of how much second and higher derivatives contribute to non-linear behavior.

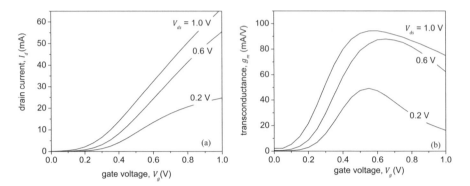

Fig. 5.2 Measured transfer characteristics of 32 nm silicon nFET. (**a**) drain current as a function of gate voltage curves at three drain voltages and (**b**) corresponding transconductance curves obtained by differentiating

Compression doesn't just reduce the output; it generates additional frequencies. This can be seen by using a power series of voltage to represent the input signal to the device and the current that it generates:

$$I\left(V_{DC}+v\right)=a_0+a_1v+a_2v^2+a_3v^3+\ldots \tag{5.1}$$

where v is the small sinusoidal voltage applied to the input of the device. The coefficients a_i depend on the applied bias voltages. In the case of an FET, I is the drain current, v is the gate voltage, and if the quadratic and higher terms can be neglected, a_1 is the transconductance g_m. Generation of extra tones comes from considering the application of a single sinusoidal tone:

$$v=S\cos(2\pi ft)=S\cos(\omega t) \tag{5.2}$$

By using this tone in the power series, and using trigonometric relations, it can be seen that

$$I\left(V_{DC}+v\right)=a_0+\frac{1}{2}a_2S^2+\left(a_1+\frac{3}{4}a_3S^3\right)\cos(\omega t)$$
$$+\left(\frac{a_2S^2}{2}\right)\cos(2\omega t)+\left(\frac{a_3S^3}{4}\right)\cos(3\omega t)+\ldots \tag{5.3}$$

These additional frequencies are harmonics, that is, they are multiples of the input frequency; hence, the distortion resulting from gain compression is known as harmonic distortion. As the input is greater than the onset of compression, then the harmonic signals can be observed on a spectrum analyzer, perhaps at frequencies even lower than the 1 dB compression point. Some spectrum analyzers have on-board software to detect and analyze these harmonics to produce a harmonic distortion figure-of-merit, total harmonic distortion (THD). The creation of harmonics can also be understood qualitatively as a result of distorting the sine wave. If a signal is larger than the compression point, a sinusoidal output becomes flattened at the top and bottom of its swing. If one imagines that the top and bottom are truncated, and starting to resemble a square wave, then this would imply that, via Fourier decomposition, that additional frequencies are required to make the distorted shape of the output signal. (This is what makes overdriving the amplifiers of electric guitars give them a rich sound.) In communication circuits, harmonics can generally be removed by narrowband filtering around the fundamental frequency.

5.3 Two-Tone Non-Linearity: Third-Order Intercept

On the other hand, when a circuit or device is stimulated with two or more frequencies at the same time, if the circuit or device is linear, the output should contain only those same frequencies. Another way of saying this is that superposition of signals holds for linear networks. Non-linearity results, though, in a variety of frequencies which are not multiples of the inputs, that is, superposition fails. Cross-multiplication terms which are sums and differences of multiples of the input frequencies occur in the output spectrum. The presence of several frequencies is likely in communication circuits which multiplex several signals together. Two-tone measurements are used to quantify non-linearity as applied to such circumstances. This is a more demanding measurement than gain compression but because of such applications, it yields a figure-of-merit which is often considered more useful. While application of multiple tones to an isolated single transistor is not a common problem (it is the domain of communication circuit design), the measurement of linearity by this means is useful in assessing the quality of the technology. The non-linearity determined by two-tone measurements is closely related to the measurable DC transconductance characteristic, g_m, and its derivatives.

(It should also be noted that non-linearity can also be used to advantage. The ubiquitous mixer circuit performs frequency translation from high RF frequency to a lower intermediate (IF) frequency by mixing it with a local oscillator (LO) frequency. The useful IF frequency is the difference between the RF and LO, which is usually substantially lower. This mixing can be achieved if the non-linearity is dominated by the quadratic, or square-law, behavior of the device or circuit. The undesired non-linearity discussed here, and measured with the two-tone method, is dominated by the cubic term, and is important even in a square-law mixer. A second-order intermodulation figure-of-merit is useful for mixer characterization.)

As the name suggests, in two-tone measurements, sine waves of two different frequencies, with the same amplitudes, are simultaneously applied to the circuit or device being tested [1]. Using the same power series expansion of the input, two tones, ω_1 and ω_2, where angular frequency ($2\pi f$) is used again, are now applied with the same amplitude:

$$v = S\left[\cos\left(\omega_1 t\right) + \cos\left(\omega_2 t\right)\right] \tag{5.4}$$

Expansion of this voltage in a power series as in Eq. (5.1), results in cosine terms with arguments of $n\omega_1 + m\omega_2$ and $n\omega_1 - m\omega_2$, where n and m are integers. In addition to the harmonics which occur with the gain compression of a single tone input described above, many of the frequencies found in this expansion are not harmonics of the input frequencies, but new frequencies generated from cross-products, or intermodulation products. Excluding the harmonic terms, the expansion yields

$$
\begin{aligned}
I\left(V_{\text{DC}} + v\right) = a_0 &+ a_2 S^2 \\
&+ \left[a_1 + \frac{9}{4}a_3 S^3\right]\left(\cos\omega_1 t + \cos\omega_2 t\right) \\
&+ \left[a_2 S^2\right]\left[\cos\left(\omega_1 + \omega_2\right)t + \cos\left(\omega_1 - \omega_2\right)t\right] \\
&+ \frac{3}{4}a_3 S^3 \left[\begin{array}{l}\cos\left(\omega_1 + 2\omega_2\right)t + \cos\left(\omega_1 - 2\omega_2\right)t + \\ \cos\left(2\omega_1 + \omega_2\right)t + \cos\left(2\omega_1 - \omega_2\right)t\end{array}\right]
\end{aligned}
\tag{5.5}
$$

where the frequency mixing terms are collected using the power of the signal amplitude, S.

The first few intermodulation product frequencies in the expression are illustrated in Fig. 5.3. Most of them are separated from the input frequencies, ω_1 and ω_2, but there are two terms in the cubic form which can be close to these values, ($2\omega_1 - \omega_2$) and ($2\omega_2 - \omega_1$). As ω_1 and ω_2 get close in value, these terms also get very close, as indicated in Fig. 5.3. This is the issue of concern for circuits. Generally, circuits include filters to pass only the desired frequencies falling within a specified band, such as shown by the dashed lines of Fig. 5.3, but these unwanted intermodulation (IM) products, unlike harmonic distortion frequencies, can fall within the band of the filter, if the fundamentals are close enough, leading to intermodulation distortion (IMD). If the spacing is wider, they might fall into an adjacent band. As these terms are generated by non-linearity of the device or circuit, and are of practical consequence for circuits, their measurement leads to a useful and widely recognized figure-of-merit for linearity.

From the equations it can be seen that these closely spaced frequencies depend on the third power of voltage, or of power. This is the basis of the third-order intercept measurement. Unlike gain compression measurement which requires power measurement, in two-tone measurements a spectrum analyzer is required to individually measure all of these output frequencies as well as their power. And also unlike gain compression, the result of two-tone measurements is an imaginary point, not physically realized. As a function of input power, the linear output grows linearly while these third-order IM products grow as the cube of the power, as seen in Eq. (5.5). This is illustrated in Fig. 5.4, where the powers are drawn logarithmically, e.g., in dBm. The x-axis is the input power, where both ω_1 and ω_2 have the same power level. The P_1 outputs are proportional to the input power, i.e., they have a slope of one,[1] and the P_3 outputs have a slope of three. Since they increase at different rates, the extrapolated terms hypothetically intersect at a point which is called the third-order intercept point, *TOI*. It is also known loosely as *IP3* (intercept of P_3). This is the value of input power at which at which the third-order intermodulation product would have the same output power as the power of the fundamental tone. This is a purely constructed number, as, in reality, gain compression would limit the power well before it could reach this value, but it represents a single number which characterizes linearity. High *TOI* is a sign of good linearity: a very linear device has small third-order coefficient a_3, hence a lower IM curve, and therefore the intercept occurs at a higher value. Usually, *TOI* is referred to the input power, where it is known as *IIP3*, or input-referred *IP3*, as shown on the horizontal axis of Fig. 5.4, but it can also be referred to the output power, called *OIP3*, or output-referred *IP3*, as shown. Since output power depends on the load which the device or circuit is driving in a particularly circumstance, *IIP3* is a more fundamental figure-of-merit than *OIP3*. Conventionally, the term *IP3* is considered the same as *IIP3*, unless a distinction is important.

Two-tone measurements are performed with apparatus, shown in Fig. 5.5, not very different from compression measurements. All of the connections are made with coaxial cables. An additional signal generator is required, and the two signals are combined into one with a combiner, which is a passive device which adds the inputs together into one output, and works over a specified bandwidth. Sometimes isolators are placed between the signal generators and the combiner to make sure that no unwanted signals get added in. The combined two-tone signal is applied to the input of the device, which in this case is the gate of the transistor, through the usual microwave probes, and the output is observed on a spectrum analyzer. Many modern spectrum analyzers have a built-in feature to measure the third-order intercept with the press of a button, but the measurements can be done easily with older, traditional spectrum analyzers. Even these new spectrum analyzers have a single input, so the user must combine the two input tones.

An external or internal attenuator may be required on the input to the spectrum analyzer because a spectrum analyzer itself exhibits some non-linearity which could be mistaken for that of the device being evaluated. Its non-linearity comes from its mixer, which can have a fairly low *IIP3*, but an attenuator can be used to reduce the signal to the mixer to avoid this problem. But attenuating effectively lowers the signal-to-noise ratio of all measurements, and the attenuator value must be chosen with some care. Some modern spectrum analyzers with built-in *TOI* capability take care of this range setting automatically. Since device measurements are performed with microwave probes, it is possible to measure the linearity of the measurement system by probing and measuring a metal "through" structure, where a metal connects the gate and drain. As metal resistors are very linear except at extreme power levels, a high *IIP3* should result. A typical spectrum analyzer should easily have an *IIP3* of greater than 40 dBm measured this way if the power level is chosen carefully.

---

[1] Note that there is a third-order amplitude, a_3 in the linear term of Eq. (5.5). If this is too large to not neglect, the analysis does not proceed, but in this case, the non-linearity is severe.

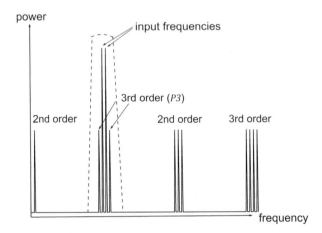

Fig. 5.3 Illustration of second- and third-order intermodulation products when a non-linear device is stimulated with two closely spaced input frequencies. The dashed line suggests a bandpass filter that might be applied in a circuit application

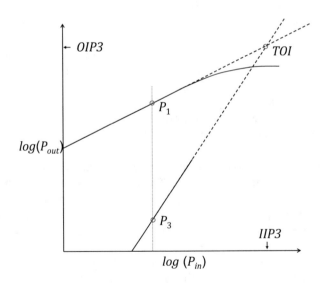

Fig. 5.4 Definition of the third-order intercept point (*TOI*) figure-of-merit to describe non-linearity

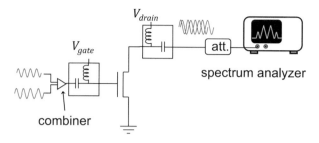

Fig. 5.5 Apparatus used to measure non-linearity with the two-tone method. Coaxial cables are used for the connections

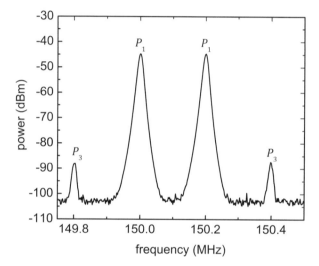

Fig. 5.6 Measured output frequencies of a non-linear device subjected to two input frequencies

It is noted that attenuating the input signal, as recommended, does not change *IIP3*, since both *P1* and *P3* signals are equally attenuated, changing the vertical, output power scale, but not changing the horizontal third-order intercept point. However, the input signal loss in the combiner and isolators, if used, must be accounted for. Attenuation in the combiner reduces the *input* power below its nominal value, which shifts measured curves to the left, and changes the value of the projection of *TOI* to obtain *IIP3*. This loss must be measured and added back to the apparent *IIP3*.

An example of an output spectrum of a typical FET subject to two-tone stimulation is shown in Fig. 5.6 [3]. The two input frequencies have the same amplitude, as do the outputs. These amplitudes should be verified before doing measurements. Any frequency can be used for the measurement, but small $f_2 - f_1$ frequency differences should be used so that the output frequencies are not reduced by bandwidth of the transistor. In this example, they differ only by 200 kHz. The resolution bandwidth of the spectrum analyzer must be low enough to clearly resolve the four

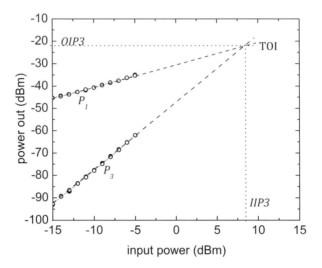

Fig. 5.7 Measured example of the determination of *TOI* and *IP3* by extrapolation

signals. The value of the fundamentals is also generally chosen to be low, for the same reason, but in a novel device it is reasonable to test at different frequencies. The input powers are swept to generate a plot like the illustration of Fig. 5.4, but note that the power must be less than the compression point, as applying power above this value will cause the fundamental output power to decrease and appear in harmonic frequencies. In other words, the *IIP3* measurements should be limited to the small-signal regime as determined from single-tone measurement. Of course, gain compression can also be observed during two-tone measurements.

An example of actual two-tone measurements and analysis is shown in Fig. 5.7, along with the extrapolation of *IIP3* and *OIP3* [3]. Note that there are four measurements for each input power, two fundamentals, and two IMD products, and that the *P3* measurements show some measurement errors, which should be minimized by averaging. The *P1* outputs have a slope of one and *P3* have a slope of three, on this logarithmic scale, as expected from the analysis above. Therefore, *IIP3* can be estimated by extrapolation, but it is also possible to obtain *IIP3* on a point-by-point basis.

Using simple geometry with Fig. 5.4, it is easy to see that for a given input power, P_{in},

$$IIP_3 = P_{in} + \frac{P_1 - P_3}{2} \tag{5.6}$$

where the power must be expressed logarithmically. Although it is advised to measure for a range of input power, to assure that linear and intermodulation terms have the correct slope, this geometric method is simpler than extrapolating to find the intercepts. An example calculation, using different data, is shown in Fig. 5.8, which shows *P1, P3*, and *IIP3 vs* $P_{in,}$ computed as in Eq. (5.6), at each value of P_{in}, as well

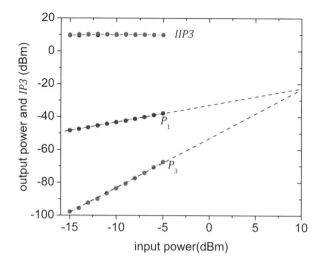

Fig. 5.8 Measured example of the determination of *IIP3* by point-to-point geometric method

as the extrapolated value. Over the input range of 10 dB, the same value of *IIP3* is obtained and agrees with the value obtained by extrapolation.

The connection between DC characteristics and this linearity figure-of-merit is made through the current dependence of voltage. Inasmuch as the linear term of the power series of voltage is proportional to the transconductance, which is the first derivative of the current, the cubic term is proportional to the second derivative of transconductance, g_m'', and *IIP3* can be related to their ratio:

$$IIP3 = \frac{4}{3}\left|\frac{g_m}{g_m''}\right| \tag{5.7}$$

where $g_m'' = \mathrm{d}^3 I_\mathrm{d} / \mathrm{d}v_\mathrm{g}^3$.

Since the third-order intercept is related to the transconductance of a device and its second derivative, these DC-measured characteristics can be used in principle, to predict *IIP3*, but in practice this is nearly impossible. The denominator term in the equation is the third derivative of the measured $I_\mathrm{d}-V_\mathrm{g}$ characteristic. Taking the third derivative of a measured quantity numerically, that is, by subtraction and division of discrete points, is a problematic, because there is bound to be some measurement error, or noise. A single measurement error propagates to points on either side of it each time a derivative is taken, and after three derivatives, significant errors accumulate. Figure 5.9 illustrates this problem. An almost unnoticeable error at $X = 15$ in an otherwise perfect quadratic function creates a large and widespread errors in the numerically determined third derivative, which should have a value of zero. The situation might be much worse when all of the points have measurement errors. Nor is it possible to fit a curve and differentiate it analytically, unless the analytical form

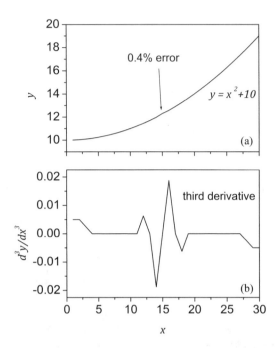

Fig. 5.9 Demonstration of the difficulty of taking third derivatives of measured data. (**a**) A quadratic curve with an error at a single point and (**b**) the third derivative of the quadratic curve determined by numerical differentiation

is a *perfect* physical description. Although an analytical expression can be differentiated without concern with measurement error, if the function is not perfect, its derivatives will be wrong.

However, it is useful to look at I_d–V_g characteristic to determine where *IIP3* is valid and to gain insight into the operation of the transistor. For example, if there is strong curvature in the I_d–V_g curve, it might be expected that its third derivative could change sign in that vicinity, causing a zero in the denominator and a spike in the third-order intercept. But a very flat g_m might also suggest low g_m'', leading to a high value of *TOI* over a wide range.

An example of measured *IIP3 vs* gate voltage of a CMOS nFET is shown in Fig. 5.10 for two values of drain voltage. These measurements correspond to the DC transfer curves shown in Fig. 5.2. There are clear *IIP3* peaks for both curves near-threshold voltage, where the curves are clearly not straight, and the shift in curves seen in Fig. 5.2(a) is apparent in Fig. 5.10. As argued above, these peaks are the result of g_m'' crossing through zero. The peak in the V_{ds} = 0.6 V curves undoubtedly has the same cause though it is not very apparent in the DC curves. Nonetheless, it is obvious that *narrow* peaks in *IP3* are not to be regarded as useful points of linearity in comparing technologies.

For comparison, a *TOI* measurement of a graphene FET is shown in Fig. 5.11 [3]. In this case, a dip is observed near a gate voltage of 0.0 V, but overall, the graphene

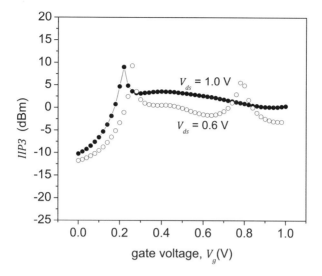

Fig. 5.10 Measured *IIP3* of a silicon nFET as a function of gate voltage, at two different values of drain voltage. These measurements are made on the same transistor whose transfer characteristics are shown in Fig. 5.2

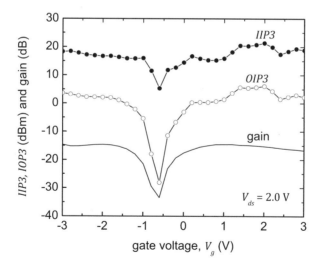

Fig. 5.11 Measured values of gain, *IIP3* and *OIP3* as a function of gate voltage, for a graphene FET

transistor shows higher linearity than the MOSFET. This is not surprising since graphene is more like a resistor than a semiconductor, and this is also reflected in the transconductance, except for the major dip. Over much of the voltage range, g_m is almost constant, so that g_m'' is very small. But near 0 V, a sharp dip corresponds to the Dirac point of the transfer curves, where because the current is very low, the transconductance drops dramatically. As graphene transistors can have positive or negative current flowing, the Dirac point is the voltage which divides these two branches, and a circuit would operate in only one branch, away from the Dirac point. Thus, if this dip is avoided, the linearity varies smoothly over a wide voltage range. It is also seen that *OIP3* is less than *IIP3* for this transistor. This is because, in this situation, where the transistor has a 50 Ω load (due to the spectrum analyzer), it actually has negative gain, as seen in Fig. 5.11. The combination of IMD measurements and transconductance can be very useful for determining where to set biases of a novel transistor in order to achieve maximum linearity, while avoiding anomalously high values over narrow bias ranges, which are the result of structure in the derivatives of g_m.

References

1. T.H. Lee, *The Design of CMOS Radio-Frequency Integrated Circuits*, Cambridge: Cambridge University Press, 1998.
2. S.-J. Han, K. A. Jenkins, A. Valdes-Garcia, A. D. Franklin, A. A. Bol, and W. Haensch, "High-Frequency Graphene Voltage Amplifier," *Nanoletters*, vol. 11, no. 9, pp. 3690-3693, Aug. 2011. https://doi.org/10.1021/nl2016637.
3. K. A. Jenkins, D. B. Farmer , S.-J. Han, C. Dimitrakopoulos, S. Oida, A. Valdes-Garcia, "Linearity of graphene field-effect transistors," *Appl. Phys. Lett.*, vol. 103, pp. 173115-1173118, 2013. https://doi.org/10.1063/1.4826932

Chapter 6
Measurement of the Large-Signal Propagation Delay of Single Transistors

Operation of transistors as digital processing components leads to specific large-signal characterization needs and measurement techniques. The primary factor which ultimately imposes a limit on how fast a circuit can operate is the device propagation delay: how long it takes for an input signal to change the output of the transistor. In digital, or logical, operation, when the static states of the transistors are in only one of two logical states (0 and 1, or low and high), the inputs and outputs are, by definition, large-signal in their static state. To assess the potential digital circuit performance of a novel transistor, it is necessary to measure its propagation delay.

This chapter presents techniques to measure large-signal propagation delay. After describing a well-known single transistor figure-of-merit, the measurement of ring oscillator frequency is used as the primary means to measure propagation delay. While ring oscillators are commonly used to evaluate conventional microelectronics, modification of the structures and measurement techniques can be used to go beyond performance measurement to assess aging, drift, and time-dependence, which might occur in novel transistors.

6.1 Single Transistor Figure-of-Merit

Digital computers and communication circuits use the concept of logical signals: a logic gate is computer element which has only two stable states, known as 1 and 0, or ON and OFF. In CMOS technology, the predominant logic gates are composed of an nFET and pFET, and the logic states correspond to the one of the FETs being on (the channel is conducting) and the other being off.[1] In CMOS logic, the output voltages swing from rail to rail, i.e., from ground to the power supply voltage and

---

[1] Differential logic gates, such as CML (current-mode logic) are often used in communication channels. These gates operate on principles quite different from CMOS, and generally use voltage swings which are smaller than rail to rail.

© Springer Nature Switzerland AG 2022
K. A. Jenkins, *RF and Time-domain Techniques for Evaluating Novel Semiconductor Transistors*, https://doi.org/10.1007/978-3-030-77775-3_6

vice-versa, with only a short time for the transition between these states. Hence, logic gates are commonly referred to as switches. The transition time between states determines the speed at which a digital circuit can operate. Just as the predominant performance figure-of-merit for small-signal operation of transistors is cutoff frequency, $f_T = g_m/2\pi C$, there is a common need to define and measure an equivalent figure for transistors used as digital switches.

Unlike the small-signal evaluation of transistors which has been discussed in the preceding chapters, operation of transistors as digital elements requires large-signal evaluation. This is not just a difference in signal size; it is a fundamentally different operation. In small-signal operation, the gate and drain of an FET are set to some voltages, and, usually, a small CW signal is applied to the gate, and is amplified. The transistor remains on and the drain current remains constant except for the small-signal perturbation. In large-signal, digital operation, the gate voltage swings between ground and V_{DD}, the power supply voltage, and the device channel switches between conducting and isolating, and remains in that state for a relatively long time. There is no amplification of voltage. The opposite of CW small-signal operation, this is the domain of transient switching signals. The performance of transistors in this domain is given by how much time is required to turn the channel from off to on, or vice-versa. Suitably defined, this can lead to a propagation delay figure-of-merit useful to predict circuit performance.

A commonly used figure-of-merit, often used in technology publications, is known as intrinsic delay, t_d, where

$$t_d = \frac{CV_{DD}}{I_{on}} \tag{6.1}$$

where C is the total gate capacitance, e.g., C_{gate}, as defined in the discussion of cutoff frequency in the "Frequency Performance Figure-of-Merit and Linear Model" section of Chap. 3, V_{DD} is the power supply voltage, and I_{on} is the saturation current at that voltage. In practice, this figure-of-merit is often just called CV/I. The origin of this figure-of-merit is the desire to have a single number with which to predict the effect of dimensional scaling, and which is easy to compute from DC measurements and low-frequency capacitance measurements. The concept is simple: the time constant to charge a capacitor in series with a resistor is RC. It is not the total delay, but the number in the exponential equation of the network. It represents the time constant of one transistor driving the gate of another identical one. The voltage of the capacitor rises to 0.632 or $(1 - 1/e)$ of its final value at time RC. The *large-signal* output resistance of an FET at saturation is V/I, hence the expression above. (A more accurate delay figure-of-merit should actually be 0.69 CV/I, as that marks the 50% rise or fall time.)

However, it should be noted that intrinsic delay, as defined above, is not actually a predictive number; it is merely used to compare device technologies in a relative way. It does not actually predict the delay of a transistor, but it does show, for example, that if the current can be increased, or the voltage reduced, say, by a fabrication change, then the delay in a digital circuit can be expected to be reduced.

A comparison with cutoff frequency is of interest. The inverse of angular cutoff frequency (where $\omega = 2\pi f$) is also a delay, and, of course, has similar dimensions:

$$t_T = \frac{1}{\omega_T} = \frac{C}{g_m} \tag{6.2}$$

But there are several differences. Of course, cutoff frequency is a small-signal concept and t_d is large-signal. Whereas small-signal implies a linear network where components have fixed values, under large-signal operation, currents and capacitances are non-linear functions of voltage, and using constants in the expression for t_d is much oversimplified. In the case of a changing voltage, the time constant (for a rising voltage) can be computed by integrating differential time constants over the voltage [1]:

$$t_d = \int_0^V \frac{C(V)}{I(V)} dV \tag{6.3}$$

Such a calculation can only be done numerically, and may not be worth the bother, and in any case, it cannot be done until a reasonably good electrical model of the device behavior has been constructed. Furthermore, this calculation represents an unlikely physical situation, where a single FET drives the gate (the load) of an identical FET. Such a condition does not occur in real circuits. In contrast, cutoff frequency is the intrinsic response of a single device, that is, it is related to the transit time of a signal from input to output of a single FET. Also, cutoff frequency uses the situation of a short-circuit output, unlike the capacitive load just mentioned.

6.2   Direct Propagation Delay Measurement

Intrinsic delay is not well defined, is not accurate, is not physically realistic, and is not the large-signal equivalent of cutoff frequency. (To be fair, it must also be pointed out that cutoff frequency is not physically realizable, but it is well defined: if cutoff frequency is known, then the short-circuit current gain can be calculated at any frequency. Bipolar transistors, in addition, require knowledge of corner frequency. And in both cases, cutoff frequency is related to the physical transit time across the device.) In spite of this, the CV/I metric is often cited in technology papers, partly because it is easy to measure and it produces numbers in the psec range, which sound attractively like gate delay. But some authors actually present the reciprocal of cutoff frequency as if it is the same figure-of-merit.

It might be thought that a better propagation figure-of-merit could be obtained by measurement methods analogous to frequency methods. One could apply a rail-to-rail signal to the gate of the transistor, and observe the output on a time-domain measuring instrument, *e.g.*, on an oscilloscope. The equipment setup for such a measurement would be as shown in Fig. 6.1. This is similar to the frequency measurements, except that there is no bias tee needed on the input side, as the full rail-to-rail voltage swing signal would be provided by a pulse generator. The output side still has to have a bias tee to apply voltage to the drain, but transient switching of the output is transmitted to the oscilloscope. All of the connections are made with coaxial cables.

In addition, a reference path must be built alongside the transistor. This is roughly analogous to de-embedding pads used for S-parameter measurements. As shown in Fig. 6.2, a "through" structure is needed. The pulse is applied separately to both the through and the transistor, by moving the probes between the transistor and the through so that the delay of the signals due to long cable paths is common to both structures, and the difference between the through and transistor signals is attributed to the transistor propagation delay. In order to display this difference, the oscilloscope must be externally triggered, using a separate channel, from an additional synchronized signal coming from the pulse generator.

This seemingly simple measurement, however, has several problems which may prevent it from being very useful. The delay to be measured is typically, in contemporary CMOS, less than 10 ps, and accuracy demands that it must be measured with some fraction of that, so 1 ps or less is a desirable resolution. Ideally, the gate would have a perfect step from 0 to V_{DD}, or from V_{DD} to 0, applied to it, that is, a purely digital input, so its output would represent simply the transistor delay. But the rise or fall time of any pulse from commercial equipment is much longer than the typical propagation delay. The shortest risetime readily available is that of the step generator in a TDR (time-domain reflectometer) option of an oscilloscope, which is typically 25–35 ps. Furthermore, it has a fixed amplitude, typically 0.20 V, so it doesn't present a full or adjustable voltage swing to the transistor's input. Some other very high performance pulse generators with adjustable voltage levels have risetimes of 60 ps. One-step generator produces a −9 V pulse of 15 ps, but after passing it through an inverter, an attenuator, and a bias tee, the risetime inevitably degrades to about 25 ps. Such a large risetime makes measurement of picosecond differences between oscilloscope traces quite difficult.

Furthermore, the pulse coming out of the drain is inverted and reduced in amplitude, as shown in Fig. 6.3 for a hypothetical input pulse of 48 ps risetime. Before the pulse arrives, the output is at 0 V because the input resistance of the oscilloscope

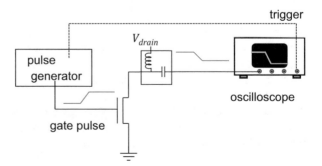

Fig. 6.1 Apparatus to measure the response of a transistor to a transient input voltage

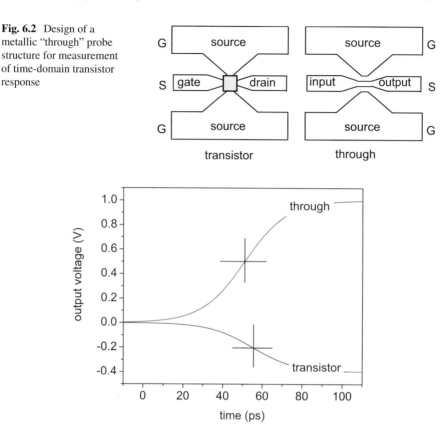

Fig. 6.2 Design of a metallic "through" probe structure for measurement of time-domain transistor response

Fig. 6.3 Illustration of the expected measurement of transistor output and through structure

(50 Ω) discharges the capacitor of the bias tee. When the signal arrives at the gate, a transient current which flows through the transistor is sourced by the oscilloscope. (The inductor blocks transient current.) Therefore, the voltage at the oscilloscope input drops below 0, to a level equal to the device current times 50 Ω. For an nFET, as shown, the current flows from the oscilloscope to the transistor, so the voltage is negative. Eventually, after a relatively long time, the voltage returns to zero, depending on the value of the capacitor in the bias tee. In the case of a very small transistor, this transient voltage may be invisibly small on an oscilloscope, but in any case, the situation illustrated in Fig. 6.3 is far from the ideal of a step function response. In fact, the illustration also neglects the effect of threshold voltage, which would cause a different shape of the output. So although it might be possible to read off the delays of the transistor with respect to the gate pulse, as shown with the crosshairs, it is an imprecise method of measuring propagation delays of single transistors when the delay is much smaller than the rise and fall times. As a practical matter, the determination of the 50% point adds to the imprecision, but determining its value from a much coarser time scale display can help. Such a measurement method might be useful for measuring delay of relatively slow transistors which are not

attempting to compete with modern silicon FETs. It can also be used for examining difference due to process modifications. For example, such a measurement was used to reveal anomalous delays, such as that caused by the high gate resistance described in Chap. 4.

6.3   Ring Oscillators to Measure Delay

Since the problem in using the direct propagation delay measurement just described is that of generating a sharply rising gate signal, it might be thought that creating a fast pulse and measuring short delays could be addressed by combining many transistors in series, on the wafer. The first device could stimulate the next one with a fast rising or falling pulse, unaffected by cables or wiring, and the remaining stages would act as to multiply the delay per stage by the number of stages in the chain, thereby producing a larger net delay which is easier to measure. Although this is not so easy as stated, since signal inversion is required for each stage, it is the general idea behind the use of ring oscillators and delay chains for time-domain performance measurement.

In publications describing advances in transistor technology, particularly CMOS, several performance figures-of-merit are usually cited. In addition to showing large numbers of DC characteristic, the publications usually report the CV/I metric, cutoff frequency, f_T, and some number of ring oscillator frequency measurements.

Ring oscillators are the predominant structures used to characterize performance of transistors in the digital, or logic, domain [2]. There are two reasons for this. First, fabrication of a ring oscillator is a technological first step toward building an integrated circuit. It requires fabricating a number of more-or-less identical, and working, transistors and connecting them together in a prototype of the integrated circuit fabrication technology. Second, they are easily characterized by measuring their frequency. Frequency measurement is far simpler and easier than any other performance measurement. For the purposes of measuring the signal propagation delay, ring oscillators can be regarded as delay multipliers, as the delay of many identical transistors is easier to measure than the delay of a single one. The same is true of timing jitter, if it is of interest. On the other hand, variability of delay between nominally identical transistors in the ring oscillator cannot be seen, as it is hidden by the fact of averaging. The question of uniformity, which is very important for digital circuit design, is therefore completely ignored. For several other reasons, ring oscillators are not ideal for measuring transistor digital speed, but there is essentially no other electrical measurement technique available.

The concept of a ring oscillator is very well-known. A series of inverting elements, or stages, are connected together to form a ring, as in Fig. 6.4(a). Each stage inverts its input and drives the next stage with the inverted signal. The simplest element is the inverter, as illustrated for complementary logic consisting of a single pFET and single nFET, but all other CMOS logic elements invert. The final stage of the ring, as in the drawing, is connected to the first stage, forming a ring of inverters.

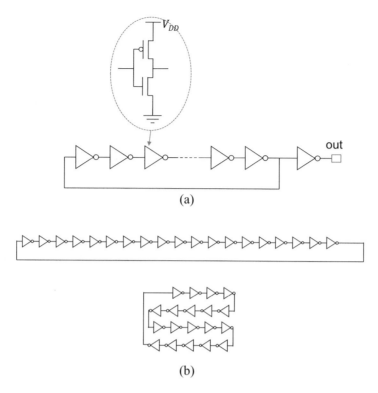

(a)

(b)

Fig. 6.4 (**a**) Schematic of a simple ring oscillator, and (**b**) two illustrations of the construction of feedback wiring

If the number of stages is odd, the ring will oscillate, as will be explained below. (If the elements of the ring are differential, a ring oscillator can be made with an even number of stages by flipping the differential inputs to one stage.) The "ring" is sometimes folded into small sections to reduce wiring capacitance, as illustrated in Fig. 6.4 (b). An important aspect of this structure is the off-chip driver device, which is connected to some part of the ring and drives an output pad which can be used to measure the frequency of the ring.

The oscillation results from the inversion at each stage, as illustrated in Fig. 6.5. The output of each stage of a three-stage ring oscillator is shown. As stage 1 switches from low to high, at approximately its midpoint, it starts a transition from high to low of stage 2. The midpoint of stage 2 starts the transition from low to high of stage 3, and so on. The time delay from the midpoint of one stage to the midpoint of the next is called the stage delay, t_p, which is the desired stage speed measurement. Rising and falling delays may not be the same, but are represented here as the average value, for convenience. After stage 3 switches, it then switches stage 1 because of the inversion at each stage. After two passes around the ring, the signals start to repeat, thus the period of oscillation, T, is equal to twice the number, n, of single stage delays in the ring:

Fig. 6.5 Illustration of the
stage-to-stage transitions
which cause oscillation in
a ring of and odd number
of inverters

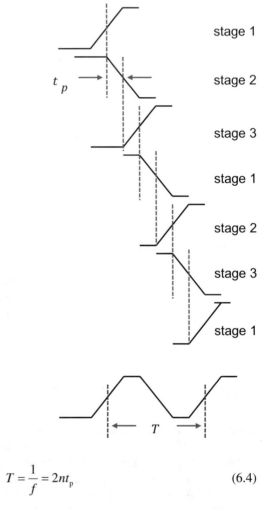

$$T = \frac{1}{f} = 2nt_p \tag{6.4}$$

Measurement of the period, which is usually actually measured by its frequency, measures the average propagation delay, also called switching delay, of a single stage. Where does the first transition come from? What starts the oscillation? Thermal noise is usually enough to cause minor variations of the output voltage, and because inverters actually have very high gain (they switch rail to rail for a fairly small change of input voltage) at each stage, a full rail-to-rail oscillation quickly develops. It is also possible to initiate oscillation with a control gate inserted into the ring oscillator, which will be described below.

The stage-to-stage inversion is fundamental to propagate a signal. The inverting elements of a ring oscillator are usually not single nFETs or pFETs although early in the history of MOS technology, logic gates were made with only nFETs or only pFETs, using resistors for pull-down or pull-up loads. Today, most logic elements are made from complementary circuits using nFETs and pFETs such as the inverter shown in Fig. 6.4.

How is a ring oscillator to be made with novel devices? Indeed, this can be quite a challenge. Three things are needed: (1) both *n*- and *p*-type FETs, or one type and a resistor, (2) sufficient device yield and uniformity to make a circuit, and (3) a wiring and integration technology sophisticated enough to fabricate a circuit. If these objectives can be met, then this small circuit can be useful in measuring the performance of the transistors as digital switches, if interpreted correctly.

How many elements should be in the ring? In mature technologies, quick and easy measurements are needed to monitor the processing quality and uniformity during manufacturing, and ring oscillator structures are designed to be measured with low-frequency outputs to avoid the issues of high frequency measurements, so the difficulties taken up in this book are unimportant. This is achieved by using a lot of stages, say, over 100, and inserting a frequency dividing circuit between the ring and the output measurement point to reduce the measured frequency to a few dozen MHz or less. A frequency divider is a relatively sophisticated and complex circuit requiring a large number of transistors. If the devices to be studied are the results of a modification of a well-established fabrication technology, this is a reasonable path. But this is often impractical for novel and exploratory transistors, and the best scenario is to take the opposite approach, and use as few stages as possible.

However, the number of stages cannot be too small. One stage won't work because the output and input are connected to each other, and inverted, but as a delay between output and input is needed to cause oscillation, a direct connection doesn't do anything. Three stages can work, as there is a loop-back delay, but it may not result in a full rail-to-rail swing. Figure 6.6 illustrates this problem. In this figure, the output of each stage is assumed to have an exponential time-dependence. Although this is only an approximation, it is a reasonably good one for this purpose.

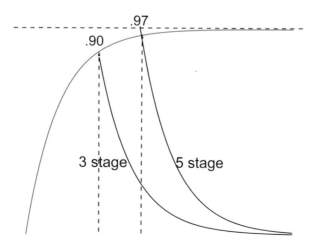

Fig. 6.6 Calculation of the rising edge of the signal from one inverter driving the input of a succeeding stage

Using the exponential approximation, the figure shows the output of a stage after one round trip, i.e., when the signal has been delayed by nt_p, where n is the number of stages, and is being pulled down by the inversion. (This is a careful drawing of the first and second stage 1 signals in Fig. 6.5.) In the three-stage ring oscillator, the first signal has only reached about 90% of the maximum voltage when it starts to pull down, so the oscillation is established at much less than full rail voltage. The five-stage ring oscillator gets to about 97% of the maximum voltage, which is fairly close to full rail. Although not shown in Fig. 6.6, the swing to ground has the same problem, failing to fall all the way to zero if there are too few stages. Of course, the real signals will be somewhat different from this idealized exponential, but it is clear that three-stage oscillators should not be used, as the logic does definitely not swing from rail to rail. If full voltage is not reached, then the propagation delay will be diminished, leading to an underestimate of the performance of the transistors. Also, the power consumption, if it is measured, will be misleading. Five stages should be the absolute minimum to be considered for a good evaluation, and seven stages are even better.

6.3.1   Ring Oscillator Frequency Measurement

An additional concern arises when the transistor yield is low, as is often the case with novel transistors. The output driver, as illustrated in Fig. 6.4, should be able to drive the measurement equipment. Because of the constraints above, that is, few stages and no frequency divider, the frequency of the ring oscillator might be fairly high, requiring high-bandwidth measuring equipment, and invoking transmission line effects in the coaxial connections. A 50 Ω oscilloscope can be used for measuring frequency, if the output transistors can provide enough current to produce a signal large enough to trigger the oscilloscope and lead to a measurement. With typical noise floors of oscilloscopes being slightly below 1 mV, a current of 1 mA is adequate, but 1 μA is far too small. Output current can be increased by using multi-finger transistors, but in an exploratory technology, this may not be possible. An amplifier can be inserted between the ring oscillator and the 50 Ω oscilloscope, but it must have sufficient gain and frequency response to boost the ring oscillator output signal.

It might be thought that an oscilloscope with a high input impedance (1 MΩ) could be used instead since the voltage at the input will not be reduced by the 50 Ω load. A high impedance frequency counter could also be considered. However, because of transmission line effects, reflections cause a significant problem. The problem is illustrated in Fig. 6.7. The output stage of the ring oscillator is connected to the high impedance oscilloscope. The signal which would be seen in a 50 Ω channel, exaggerated in amplitude, is shown for reference in the bottom trace. As a result of the high impedance, essentially an open transmission line, the signal arriving at the oscilloscope is reflected back to the ring oscillator output stage. But since this output is not designed for impedance matching (unlike test equipment), the signal

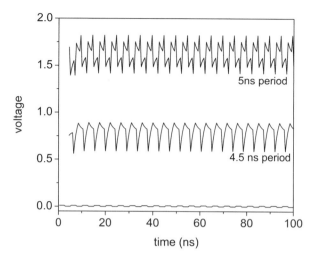

Fig. 6.7 Calculated traces which would be seen on a high impedance oscilloscope driven by the output of a ring oscillator

will get reflected again, and travel back to the oscilloscope, where it will be reflected again, and so on. Figure 6.7 shows what happens when there are three such reflections and a 20% attenuation at each reflection. The reflected waves superimpose to form a wave which depends on frequency and length of the cable. While this appears to be a wave which repeats with the correct frequencies, in reality it is not measurable. A problem arises when the frequency changes. As seen in Fig. 6.7, two different frequencies give very different waveforms, with large shape and amplitude variations. In a real, physical, ring oscillator, a slight variation in frequency, which is unavoidable due to jitter, results in this kind of output signal jumping through various waveform shapes and levels, making for an unstable, and unusable, trace on the oscilloscope. Potentially, a single triggered trace of a real-time oscilloscope could be used to extract the period. A frequency counter, which typically averages signals over a window of 1 msec or more, cannot be used at all.

A possible solution to this problem is to use a high impedance active amplifying probe, as described in Chap. 4, placed on the output pad which is *not* contacted with a transmission line, and connected to an oscilloscope. Because there is no bias tee required in this situation, there is no transmission line, either, and the complexities shown in Chap. 4 will not occur.

A far simpler method is to use a spectrum analyzer to measure the frequency. In this case, the output stage of the ring oscillator is connected directly to a spectrum analyzer, with a coaxial cable. The spectrum analyzer is a 50 Ω instrument, so the signal driven by the ring oscillator output will be absorbed by the spectrum analyzer, and none of the reflections suggested in Fig. 6.7 will occur. Because of the high sensitivity, of the spectrum analyzer, a signal which is too small to see in a 50 Ω oscilloscope can be easily detected. Furthermore, microwave probes may not be needed if the output signal is large enough so enough of it is transmitted through the inductance of the low-frequency probes that it can be detected with the spectrum analyzer. Spectra like those illustrated in Fig. 6.8 will be seen [3]. As the power supply voltage is changed, the frequency of the oscillator will change, and if the output stage uses the same voltage as the oscillator, its amplitude will change. It is also possible, and good practice, to use a separate power supply for the output stage. In this case, the spectrum analyzer will show the same amplitude for all frequencies, as in the dashed lines.

6.3.2   Ring Oscillators with Novel Transistors

Just what does the frequency of a ring oscillator measure? As discussed above, transistors can be characterized in the frequency-domain by their response to a stimulus, resulting in the cutoff frequency figure-of-merit, which measures the response of the device in isolation. To see how it responds when it drives a load, there are other figures-of-merit that can be derived from S-parameters, or it can be simulated in that situation using the linear network obtained from the S-parameter measurements. In the digital, large-signal, time-domain, there is no equivalent two-step process. There is no intrinsic self-propagation in the absence of a load, although the CV/I metric gives the appearance of one. As explained above, the C in CV/I is the capacitance of another identical device. So the speed performance of a digital transistor or logic gate, which governs the maximum speed of a circuit, is fundamentally circumscribed by the load it drives. Since, as explained above, there is no practical way to measure the delay of a single transistor, ring oscillators are used to measure the delay of logic gates. In CMOS, and in nMOS and pMOS, loads are capacitive due

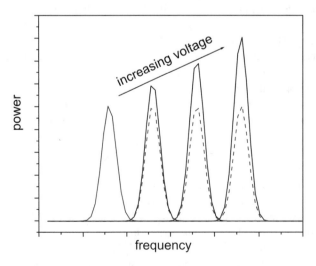

Fig. 6.8 Frequency spectrum which will be seen if spectrum analyzer is used to measure frequency of a ring oscillator measured with different voltages

to intentional connection to additional logic gates, as well as that due to parasitic wiring. Even if the input to the logic gate had an instantaneous rise or fall, the output would have a fall or rise which is approximately exponential and governed by the output resistance and *load* capacitance.

The ring oscillator, then, measures this performance. The stage-to-stage delay of a ring oscillator describes how quickly one stage, a logic gate, drives the next stage, and the load which that stage presents is fundamental to the propagation delay figure-of-merit. In conventional CMOS technology, ring oscillators are fabricated with various fan-out loads, where fan-out is the number of identical logic gates which a gate drives, or the equivalent capacitance. If the stage is an inverter, then a fan-out of three, for example, means that each inverter drives three inverters, as illustrated in Fig. 6.9. The inverters indicated by the bracket are "dummy" devices, that is, they have no function, other than to act as loads for the inverters in the oscillation path. They may or may not be connected to power supplies. The capacitive load is thereby three times that of a single inverter, such as is portrayed in Fig. 6.4. Measuring a delay with various loads is useful for developing a model for use in simulators, and for giving an idea of what performance can be achieved in real logic circuits, where fan-out is always greater than one.

When used to evaluate exploratory transistor technologies, it is important to understand the effect of loading between gates. Often a novel transistor has very low current, and, as such, is slow to drive its load capacitance to the rail voltage. As a result, the ring oscillator is often dominated by the interconnection between stages and not by the fundamental switching speed of the transistors. For example, in [4] a ring oscillator was made from a single long carbon nanotube, with multiple nFETs and pFETs fabricated on the same nanotube. While this was a technical *tour-de-force* which showed the feasibility of making logic gates with carbon nanotubes, the capacitance of the relatively very large wires between stages needed to make this circuit so overwhelmed the drive current that the frequency achieved was very low. If the interstage wiring load severely limits the ring oscillator frequency, then nothing much can be learned about the fundamental propagation delay of the transistors.

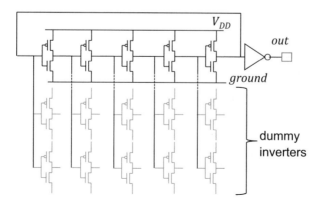

Fig. 6.9 Schematic of a ring oscillator which has dummy load stages to simulate circuit fan-out

If the transistor fabrication precludes making multistage ring oscillators and output drivers, the possibility of a hybrid technology can be considered. If the new device results from the fairly simple deposition of a new channel material, such as two-dimensional crystal, rather than the result of an elaborate technology process, such a hybrid is an interesting possibility. In this case, a short silicon CMOS ring oscillator can be used as a test bed. Instead of completing the loop with conventional FETs, a ring is fabricated with a missing stage, or a missing single FET in an inverter, and the new transistors, or transistor, are used in the empty place. This is, of course, a fabrication challenge with no simple prescription, but the demonstration of buried gate graphene FETs, described in Chaps. 3 and 4, is encouraging [5]. Rather than depositing gate, source, and drain contacts on the top side of the graphene, the gate was embedded in the substrate, the graphene placed on top, and the source and drain contacts added after that. Although this was demonstrated in a stand-alone wiring-only substrate, it is conceivable that the substrate could consist of simple CMOS transistors and interconnects, allowing for the hybrid circuit. For testing, a completed CMOS reference ring oscillator would also be fabricated, and the difference in frequency between the reference and the hybrid would be due to the new transistor. Small frequency differences are easy to measure, so the effect of replacing just one transistor out of ten (in a 5-stage inverter ring) would be easy to detect.

6.4 Instability in Digital Gates

While ring oscillators are used primarily to establish performance, i.e., stage delay, of the transistors, they are also very useful and easily used to identify time-dependent effects, such as stability. In conventional silicon technologies, the transistors are expected to be, and are, very stable when operated at nominal voltage and within a specified temperature range. Circuit design can proceed only after time-independent device models are developed, even if some margin is included for small variations. After being operated for many years, however, CMOS FETs do show various kinds of deleterious aging effects, such as bias-temperature instability (BTI) and hot-carrier injection (HCI) degradation, which are usually studied by operating them for short times at temperatures and voltages outside of the nominal range, thereby accelerating the aging. But novel transistors cannot be assumed to be stable over periods of years. In fact, new devices often show significant drifts over short periods such as minutes or seconds. One simple indication of this is the hysteresis often seen (and often judiciously omitted from publications) in I_d–V_g curves, as illustrated in Fig. 6.10. The gate is first swept from low voltage to high, as shown by the arrows, and then measured in reverse, from high to low, resulting in currents which differ due to an apparent change in threshold voltage. Similarly, a change in transconductance might be seen in an I_d–V_g curve if current flow degrades the channel mobility, with or without an accompanying change of threshold voltage. Such effects can be seen easily with the DC I–V curve, as in the example, but the changes occur over

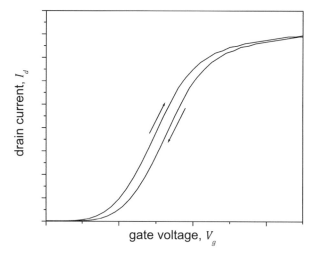

Fig. 6.10   Measured hysteresis seen in the transfer characteristic (I_d–V_g curve) of a silicon nFET

times of tens of msec or more, and considerable averaging is done for each measurement to reduce noise, which also acts to remove short-time fluctuations in output. Ring oscillators, with some modifications, can provide a closer look at stability on much shorter time scale, or under transient conditions.

Time-dependent variations of transistors' behavior can be thought of in three categories: aging, rapid fluctuations, and drift. Aging is a permanent, or semi-permanent,[2] change of transistor electrical characteristics, which depends monotonically on the time that the device is operated or has voltage applied, whereas rapid fluctuations are random variations of the characteristics which don't result in a permanent change, and drift is a slow change of characteristics which may not depend monotonically on the time of operation and may spontaneously reverse. The rapid fluctuations and drift may be largely similar, except for the time scale of the observation, but generally have different physical origins. Aging inevitably leads to a degradation, not an improvement, of useful device characteristics.

6.4.1   Gated Ring Oscillators and Delay Chains

Typically, ring oscillator frequency is measured with a frequency counter, an oscilloscope, or a spectrum analyzer. Of these, typically the shortest measurement time is achieved with a frequency counter, which can have a measurement time aperture as short as 100 or 200 μsec, though 1 msec is more typical. But if fluctuations of

---

[2] Some aging effects show partial recovery to the initial state if the operating voltage is removed for a long time.

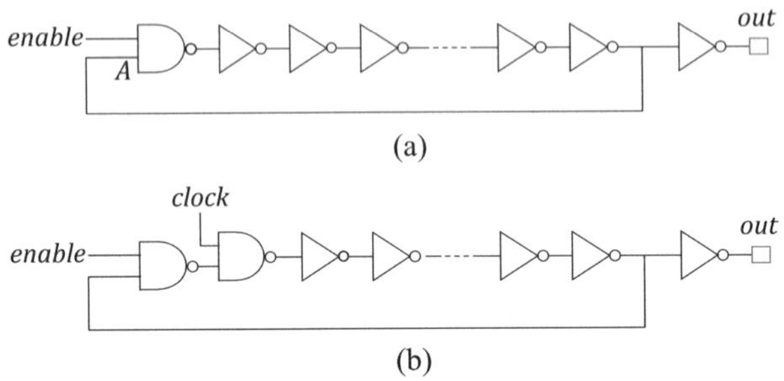

(a)

(b)

Fig. 6.11 Schematic of gated ring oscillators. (**a**) Ring oscillator with a single enable gate and (**b**) ring oscillator/delay chain with enable input and clock input

frequency occur at a shorter time scale, they will be averaged during the measurement time and not be detected. But ring oscillators can be modified to make for relatively easy observation of the effects of such variations. In Fig. 6.11, the basic ring oscillator structure is modified in two ways. In Fig. 6.11 (a), a logic NAND gate is inserted into the ring. The total number of inverting elements in the ring must still be an odd number, noting that the NAND also inverts. The effect of the NAND is to enable or disable the oscillation, by blocking the looped-back signal, making it a *gated* ring oscillator. When the enable input is zero, the signal present at input A has no effect: the output has an unchanging value of logical one, no matter what value is present at A. If the total number of stages is odd, the signal at A will be high, a logical one. If the enable is switched from low to high, the output of the NAND switches to low, and the switching of states through all of the inverters and, arriving at the NAND, switches it, too, so an oscillation is established. Just like the ungated loop, the oscillation frequency is again determined by the stage delays.

This gating technique is commonly used in conventional CMOS evaluation designs because it allows a large number of ring oscillators, which may have variations in fan-out, n/p width ratios, length, etc., to share a single common output driver, thereby simplifying measurement and saving some silicon area. Multiple ring oscillator outputs are connected to the same point, possibly through a multiplexor, or another set of NANDs, but all have individual enable signals. The ring oscillator to be measured is selected by raising its enable signal to a logical one, allowing it to oscillate. The other oscillators remain in a static condition.

6.4.2 Transistor Aging

The gated ring oscillator is much more useful than simply selecting or de-selecting this circuit: it gives any easy way to measure dynamic, time-dependent effects in simple switching circuits. As the enable signal starts the first cycle of oscillation, it provides a starting time to observe any change of the frequency (actually, the period) of the ring oscillator which occurs as a result of the switching activity. This can be particularly useful in aging studies. Even if there are fluctuations of frequency, the ideal circuit has the same average performance forever, so an ideal ring oscillator runs forever at the same frequency. New transistor technologies should be examined in this context. The experience of CMOS aging is instructive in this regard. Conventional CMOS circuits are close to the ideal, but even so, all CMOS technologies show slight degradation after long use. Gated ring oscillators can be used to advantage to study this degradation.

Two well-known aging mechanisms, bias-temperature instability (BTI) and hot-carrier injection (HCI), follow power law time-dependence,

$$D = A(V,T)t^n \qquad (6.5)$$

where D is a measure of degradation, such as an increase of threshold voltage, or a relative decrease of propagation delay, A is an amplitude which describes the magnitude of degradation, t is time, and the exponent, n, typically between 0.2 and 0.5, describes the rate of degradation. The amplitude, A, depends on voltage and temperature, and while the aging is slow under normal conditions, operating at high voltage and temperature can greatly accelerate the degradation. This power law form has a very rapid rise at short times, as illustrated in Fig. 6.12, and gradually becoming slower. One of the needs of reliability studies is to be able to predict the long-term degradation based on extrapolating from such a power law. Missing the first few moments of degradation under accelerated conditions can lead to large extrapolation errors, so there is a great need to measure degradation starting from the first application of a signal. This is known as an on-the-fly measurement, as the measurement is made without interruption to acquire data [6]. The gated oscillator is a way to do this, which particularly applies to the HCI aging mechanism.

The ring oscillator output is connected to an oscilloscope, and its trigger is fed a signal which is also used to raise the enable input to the NAND gate. The oscilloscope must be of the real-time type, so it can be set for a single triggered

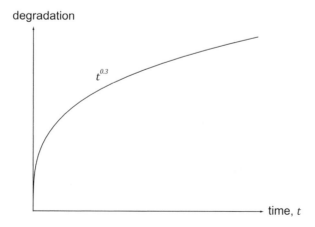

Fig. 6.12 Form of the power law degradation observed in CMOS FETs

measurement (a one-shot). That is, it responds, and digitizes, the input signal once, when it receives the trigger, and then stops responding until it is re-armed. When the enable/trigger signal is activated, the oscilloscope is triggered, and the ring oscillations are observed on the oscilloscope. As suggested by the drawing of Fig. 6.13, if there is a degradation of frequency due to the starting of switching activity, this degradation will be caught on the oscilloscope as an increase of period, and can be seen on a time scale which might be as short as a microsecond. The memory depth of modern real-time oscilloscopes can be large enough to record a significant number of cycles, after which other measurements can be initiated. A memory depth of 100,000 voltage-time measurement points is quite common. If, for example, 30 measurement points are needed to construct a cycle (the number of points in Fig. 6.13), then about 3300 cycles can be recorded for a single oscilloscope trigger. If the oscillator period is, say, about 10 ns, then this measurement would map out the first 33 μs of degradation in fine detail.

It must be pointed out that this technique works for degradation which is caused by digital switching activity, such as HCI, where substantial current flows through the channel for a short time. On the other hand, the other well-known CMOS degradation, BTI, is primarily activated by power supply voltage, and starts upon the application of that voltage and is largely independent of switching activity. The gated ring oscillator is not as valuable in this case, but the same triggering setup can be used, where the oscilloscope trigger is now the circuit voltage, which is abruptly raised to the desired level, making sure the enable input is also at the required voltage.

For a new transistor technology, if aging is observed or suspected, a gated ring oscillator makes it possible to determine if aging is caused by switching, or voltage, some combination, or due to some other factor. The gated ring oscillator makes it

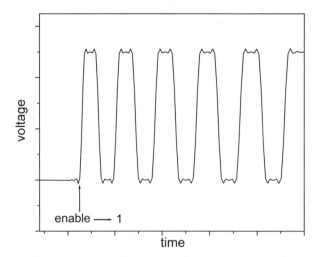

Fig. 6.13 Illustration of the slowing down of a ring oscillator, due to degradation, when powered on and activated for the first time

possible to easily see if there is a rapid or sudden change of frequency. It can also be used to see if there is a recovery after the switching has stopped or the voltage removed. Repeating the one-shot measurement after a suitable time, such as seconds or minutes, will reveal if the circuit returned to its starting frequency, or if the induced degradation became permanent.

6.4.3   Drift and Jitter

Long-term drift is also quite possible for new transistors. Rather than the one-direction degradation discussed above, it is quite common to find transistors changing their characteristics in a seemingly random manner. Ring oscillators can be easily monitored on a coarse time scale to measure such drift. Recording the frequency of the oscillator at suitable measurement intervals will reveal such drift.

But fluctuations in frequency which occur on a short time scale may be more difficult to observe. One simple method is to use the oscilloscope as in Fig. 6.13, but let the ring oscillator keep running and repeatedly record a full memory of the output, and examine the downloaded waveform memory to look for variations in the period. This can be done just as well with a simple, ungated, ring oscillator and a real-time oscilloscope. This method is closely related to one method of studying timing jitter, which is an important subject in computing and communication circuits.

Jitter can be thought of as noise in the time-domain. It is the random deviation in time of a waveform edges from their expected value and is a fact of life for all oscillators. So in the suggestion of the previous paragraph, if the average period of the ring oscillator is 10 ns, an examination of all of the rising edges might show, for example, that there is a distribution of periods ranging from 9.5 ns to 10.5 ns. This would be one measure of the waveform jitter.

To be more precise, there are two classification of jitter which might be of utility in evaluating the stability of a novel transistor technology. Figure 6.14 illustrates the two types, known as period jitter and tracking jitter. Period jitter, illustrated in Fig. 6.14(a), is the deviation of the period from its average value and is a useful figure for characterizing ring oscillators, or any other oscillator. What is shown is what might be expected to be seen on an oscilloscope which triggers repeatedly (not a one-shot trigger), and does not average the traces, instead letting each trace persist on the screen. Either a real-time or a sampling oscilloscope can be used. As indicated, the oscilloscope is triggered by the first edge of the input waveform, which is therefore always at the same position on the screen. Subsequent edges show variations in their occurrence, *i.e.*, the period varies. The variation, period jitter, gets worse as subsequent cycles are displayed, as the variation in period of one cycle becomes the starting period for the next, and jitter accumulates with increasing time.

Period jitter can also be measured with a spectrum analyzer. Since it measures frequency, a spectrum analyzer effectively measures period of a signal. A perfect oscillator, one with no jitter, would show a very narrow peak on the spectrum

Fig. 6.14 Waveforms to illustrate two forms of timing jitter. (**a**) Period jitter and (**b**) tracking jitter

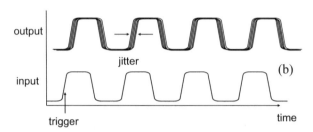

analyzer, with the width determined solely by the resolution bandwidth setting. An oscillator with period jitter, though, would have a widened distribution where the width is determined by how much variation there is in the frequency. This variation seen in the frequency-domain is called phase noise. A ring oscillator has a digital-like output, with sharp rising and falling edges, so it generates lots of harmonics, and only the fundamental frequency should be examined with a spectrum analyzer. The time-domain jitter parameter can be obtained from the phase noise measured on a spectrum analyzer, but usually phase noise measurements are used for evaluating very stable analog oscillators, such as LC tanks, where the jitter is so small as to be almost unmeasurable in the time-domain using an oscilloscope.

Tracking jitter, as in Fig. 6.14(b), is generally more appropriate for evaluating circuits, such as phase-locked loops and delay generators, where the output period follows that of the input. Delay generators are circuit elements used in communications, but a simple delay generator is an open-ended chain composed of a string of inverters, much like a ring oscillator, but not connected in a closed loop. Tracking jitter measures how accurately such a circuit's output signal follows the timing of an incoming signal, which indicates the overall jitter generated by the circuit on a repetitive basis. The oscilloscope is triggered repeatedly by the input signal, which is typically a reference clock, and jitter is measured at the output which is generated, and tracks that input. There is no spectrum analyzer measurement which is equivalent to tracking jitter. Tracking jitter can be useful for studying how new transistors or logic gates respond to various circumstances of incoming signal. For example, the hysteresis seen in DC current measurements in Fig. 6.10 might be frequency or duty-cycle dependent. At high enough frequency, the physical mechanism which causes the hysteresis, such as charge trapping, might not respond, and the current in both directions may be the same. Therefore, a high frequency incoming clock signal might generate relatively little tracking jitter in a delay chain measurement, but a lower frequency or low duty-cycle clock might generate more, as the hysteresis starts to occur. It might also reveal that the net propagation delay, absent any jitter, depends on the input frequency or duty cycle. Such effects cannot be observed with a closed-loop ring oscillator.

An example of a jitter measurement made with an oscilloscope is shown in Fig. 6.15. In this example, the triggering signal is not shown, in order to look closely at the examined edge. It is easily seen that there is variation of the arrival time of the waveform, but in addition, the oscilloscope also displays a histogram, shown in blue, of this variation of the time of the midpoint of the waveform, and computes its statistical parameters. In this case, the jitter is approximately Gaussian, with some broadening at the base. Both real-time and sampling oscilloscopes can make these measurements. As usual, the instrument itself has a lower limit, that is, an intrinsic jitter which it would measure even if the signal were perfect. Generally, sampling oscilloscopes have lower intrinsic jitter than real-time oscilloscopes, but the measurements take longer to acquire. The best oscilloscopes have intrinsic jitter of less than 1 ps rms. With a real-time oscilloscope, the waveform of interest provides the trigger, and the 0th cycle, the triggering edge, shows the instrument's trigger jitter, and subsequent cycles show period jitter.

Sampling oscilloscopes, however, need an external trigger, and it must precede the measured edge by typically 25–30 ns. Figure 6.16 illustrates this configuration. All of the connections are coaxial cables. For *period* jitter measurement, then, the signal is split by a power splitter, a resistive tee, which matches the 50 Ω impedance in all directions, and one output of the splitter is delayed by a sufficiently long coaxial cable so that its signal arrives delayed relative to the other output, which is used to trigger the oscilloscope. The cable delay can be thought of as storing the signal while the oscilloscope gets ready to measure it. If the frequency of oscillation is such that several edges are displayed on the time scale of the oscilloscope, the triggering edge can be identified as the one with the smallest jitter, which is the intrinsic jitter of the oscilloscope. Under good circumstances, this splitting and

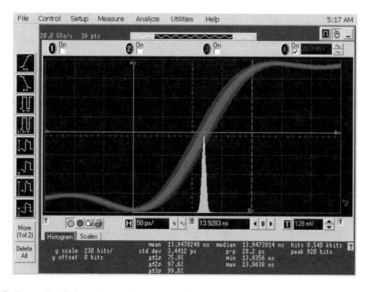

Fig. 6.15 Example of observing and measuring timing jitter by oscilloscope

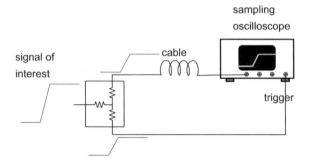

Fig. 6.16 Apparatus required to use a sampling oscilloscope to measure jitter. The connections are made with coaxial cables

triggering gives the best jitter measurement, but as it reduces the amplitude of the signal by one half, and the cable may degrade its rise and fall times, this delay may impact the measurement when the ring oscillator signal is not very strong. Although delay is needed for *tracking* jitter, power splitting the output signal is not required, as the trigger comes from the input.

The purpose of measuring jitter in a novel transistor is to determine stability and repeatability of the propagation delay. For comparison with conventional transistors, a contemporary CMOS single inverter has been shown by experiment to generate jitter as low as 3 fs per stage [7]. The jitter which accumulates in a ring oscillator, or an open-ended chain of inverters, is due to intrinsic transistor variation, transistor instability, *and* external noise sources, such as power supply or substrate noise. An external noise source causes correlated jitter, that is, all stages of the chain experience the same effect simultaneously, whereas for intrinsic transistor delay time variations, there is no correlation between stages. These two effects can be distinguished by measuring the jitter *versus* time delay, that is, the time from the trigger signal to the measured edge. Uncorrelated jitter depends on the square root of the delay time, and correlated jitter depends linearly on the time [8]. The correlated and uncorrelated jitter will be seen as two slopes on a log-log plot, as illustrated in Fig. 6.17. Naturally, a good measurement environment has low substrate noise and low power supply noise, so that the intrinsic device jitter should dominate, but it is good practice to measure to a large enough delay that the change in slope due to correlated jitter can be clearly seen. Then the intrinsic jitter can be obtained by subtraction, and compared to jitter observed with existing technologies.

Where as a simple ring oscillator is sufficient to obtain period jitter, a delay chain driven by a clock is required to measure tracking jitter. The circuit illustrated in Fig. 6.11(b) is constructed to be operated as either a free-running ring oscillator, a gated ring oscillator, or a delay chain, in order to explore the degradation, drift, and jitter of the devices in a single structure [9]. When the enable and clock inputs are both high (and the number of stages is odd), the circuit is a simple free-running ring oscillator. If the clock signal is held high, the enable NAND acts as a gate, as seen before, so the ring oscillator can be turned on or off with the enable signal. The circuit becomes a delay chain if the enable signal is set low. In this case, the inverters act as an open-ended delay chain which propagates the signal presented at the

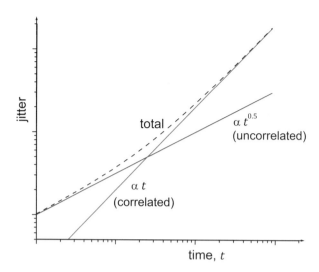

Fig. 6.17 Jitter contributions due to uncorrelated and correlated sources

clock input. As the use of the circuit in Fig. 6.11(b), illustrates, a ring oscillator is just a delay chain which loops back on itself. This makes operation very easy. But unlike a delay chain, the frequency and duty cycle of a ring oscillator is fixed. The signal activity of a delay chain, in contrast, is controlled entirely by the applied clock signal. This can be used for the measurement of tracking jitter, as described above, in order to have insight into the time-dependence of the device response. It can also be used to test how the logic gate delay of repetitive signals can differ from that of non-repetitive signals such as pseudo-random bit sequence (PRBS) patterns and data balancing, an effect which was seen in SOI technology because of its history effect [10].

The measure of jitter with a histogram results in a few statistical numbers to characterize the jitter, whether tracking or period. These numbers are the rms deviation and the peak-to-peak values. The plot of the histogram is useful for visualization, and may reveal asymmetry, but basically all of the period variations which jitter measures are swept into one graph. Like measuring voltage noise with a single number, which does not reveal the frequency spectrum of the noise, measuring a single jitter histogram does not reveal the time-dependence of the period or tracking variations. A single figure is sufficient for circuit design where a safe timing margin must be used, but it may not be adequate to understand the delay properties of a novel transistor. Slow variation, which can be called drift, of frequency can be made with repeated measurements of frequency with a frequency counter, but a jitter measurement includes drift and very short-time variations. Modern real-time oscilloscopes can be used to determine the time scale on which jitter-causing variations occur. As discussed in Sect. 6.4.2, a real-time oscilloscope can be used to trap a large number of oscillator or delay chain outputs with one-shot trigger, as in Fig. 6.13, but with many more pulses. Powerful on-board software can be used to compute the period of the measured signal, and from that, a period *vs* time plot can be extracted. The histogram is the accumulation of these periods, but the time-dependence is hidden. While the number of output edges which can be acquired with one trigger is limited by the memory depth of the oscilloscope, such an analysis can be very useful to understand the time-evolution of the transistor propagation characteristics.

In exploring the stability of novel transistors used as digital circuit elements, the use of conventional jitter measurement and analysis techniques, in combination with gated ring oscillators and delay chains, results in powerful ways to observe variations on a wide variety of time scales and input conditions.

References

1. J. M. Rabaey, *Digital Integrated Circuits: A Design Perspective*, Englewood Cliffs, NJ: Prentice Hall, 1996.
2. M. Bhushan and M.B. Ketchen, *Microelectronic Test Structures for CMOS Technology*, Switzerland: Springer, 2011.
3. J. Tang et al, "Flexible CMOS integrated circuits based on carbon nanotubes with sub-10 ns stage delays," *Nature Electronics*, vol. 1, pp. 191-196, March 2018. https://doi.org/10.1038/s41928-018-0038-8.
4. Z. Chen *et al*, "An Integrated Logic Circuit Assembled on a Single Carbon Nanotube," *Science*, vol. 311, no. 5768, March 2006, pp. 1735-1738, https://doi.org/10.1126/science.1122797.
5. S.-J. Han, K. A. Jenkins, A. Valdes-Garcia, A. D. Franklin, A. A. Bol, and W. Haensch, "High-Frequency Graphene Voltage Amplifier," *Nanoletters*, vol. 11, no. 9, pp. 3690-3693, Aug. 2011, https://doi.org/10.1021/nl2016637
6. T. Kim, R. Persaud and C. H. Kim, "Silicon Odometer: An On-Chip Reliability Monitor for Measuring Frequency Degradation of Digital Circuits," *IEEE Journal of Solid-State Circuits*, vol. 43, no. 4, pp. 874-880, April 2008, https://doi.org/10.1109/JSSC.2008.917502.
7. unpublished work of the author
8. A. Hajimiri, S. Limotyrakis and T. H. Lee, "Jitter and phase noise in ring oscillators," *IEEE Journal of Solid-State Circuits*, vol. 34, no. 6, pp. 790-804, June 1999, https://doi.org/10.1109/4.766813.
9. K. A. Jenkins, A. P. Jose and D. F. Heidel, "An on-chip jitter measurement circuit with sub-picosecond resolution," *Proceedings of the 31st European Solid-State Circuits Conference*, 2005, Grenoble, France, 2005, pp. 157-160, https://doi.org/10.1109/ESSCIR.2005.1541583.
10. K. A. Jenkins, S. Kim, S. P. Kowalczyk and D. Friedman, "Impact of SOI History Effect on Random Data Signals," *2007 IEEE International Conference on Integrated Circuit Design and Technology*, Austin, TX, 2007, pp. 1-4, https://doi.org/10.1109/ICICDT.2007.4299531.

Chapter 7
Measurement of the Transient Response of Transistors

All transistors are potentially subject to transient response effects where the electrical output of the transistor immediately after a change in its input differs from its output after that input has been held in a steady-state or DC condition for a relatively long time. Most well-established conventional semiconductor transistors have small or negligible transient effects, but novel transistors may be subject to quite large effects. This is not just an inconvenience for modeling or designing circuits. A transient response which is very different from the steady-state response might suggest that the transistor is far superior, or inferior, when used in transient operation than it is in DC operation. Digital computing circuits, it is noted, depend entirely on transient operation.

This chapter describes several techniques to measure transient response and compare it to DC behavior. The techniques rely on the application of fast edges and short pulses to the input of the transistor, and a number of methods are shown by which the response to the input can be measured on a useful time scale.

7.1 Transient Effects in Transistors

A variety of physical phenomena can change the electrical characteristics of transistors in such a way as to affect their digital propagation delays by changing their output currents. Among these *known* phenomena are:

- self-heating
- body-charging
- dielectric charging
- channel-dielectric interface charge trapping
- channel mobility instability
- material breakdown
- sensitivity to atmospheric contamination

© Springer Nature Switzerland AG 2022
K. A. Jenkins, *RF and Time-domain Techniques for Evaluating Novel Semiconductor Transistors*, https://doi.org/10.1007/978-3-030-77775-3_7

Such phenomena occur on various time scales. Except for the last two items, they can all be regarded as some form of transient effect which begins when the transistor is first turned on. This chapter presents several techniques by which to observe the transient output of single transistors, which might be the result of such phenomena, or some other. The techniques in this chapter have the goal of measuring the discrete device current on a short time scale, where a time is considered *very* short when it is comparable to the time that current flows through an FET in a digital logic gate and to compare it to current obtained at DC or relatively long times. The short time scale of interest can be less than ten picoseconds since digital circuits switch on and off at such a scale. No electrical techniques are known which can make such measurements directly, but some can reach the range of a few nanoseconds.

This goal can be considered as complementary to the studies of digital delay using ring oscillators and delay chains described in Chap. 6. Those structures measure the propagation delay which is governed by the current of the individual transistors, but as they use complementary logic, the currents flowing through the FETs of a digital gate are always transient in nature, and variations such as aging, fluctuations, and drift may arise from other sources. The concern here is also quite different from the small-signal CW operation described in earlier chapters. In those studies, a steady-state condition was *assumed* and small signals were applied; here, the question is whether immediately after the transistor is biased to the on-state, does it behave as it would in steady-state?

The need for such measurements is illustrated by the example of Fig. 7.1 [1]. Conventional manufactured CMOS transistors provide a predictable and constant current (except for small year-over-year degradation). This figure shows that while drain current of a silicon-on-insulator (SOI) FET continues to flow when the gate switches from its low to high value and is held at a constant voltage, it can change significantly after a few hundred nanoseconds from that initial change. This is a

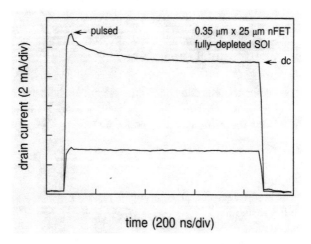

Fig. 7.1 Measured reduction of drain current of an SOI FET stimulated with a voltage step applied to its gate [1]. <© [1997] IEEE. Reprinted, with permission, from Ref [1]>

result of self-heating. Except for being fabricated on an insulating layer on the substrate, SOI FET structures are very similar to their conventional CMOS counterparts which are built on bulk Si substrates. In addition to providing electrical insulation which lowers capacitance and increases speed, the insulating layer, SiO_2, also has lower thermal conductivity than silicon, which leads to self-heating [2].

Self-heating is the heating of a transistor's conducting regions above the ambient temperature, caused by current flowing though the device. This happens to a small extent for every current conducting device, but when an FET channel is fabricated on a thermally insulating region or substrate, as it is in SOI, or with a thin-film transistor on glass, the heating can be quite severe. It is also a concern for finFETs, which have geometrically less contact with the substrate than planar FETs, and strained-silicon FETs, where the strain is induced by alloying the Si with Ge, reducing its thermal conductivity. The rise in temperature is proportional to the power in the channel times the thermal resistance, $\Delta T = P\, R_{TH}$. The thermal resistance is determined by the thermal conductivity of all of the materials surrounding the channel, but because of geometry, it is usually dominated by the layer beneath the channel. Because materials have a thermal capacitance, too, the temperature does not increase immediately, but rises exponentially with an approximately RC time constant, which is on the order of 10–100 ns in contemporary SOI devices. Temperature will always have a fairly strong effect on the transport properties in the semiconducting channel. As greater temperature results in more scattering of carriers, it leads to a reduction of the current flowing from drain to source.

In Fig. 7.1, the transistor is turned on abruptly, by activating the gate with a long voltage pulse, with a sharp rise time, from a pulse generator. Current is measured by a transformer technique, which will be described below. The upper curve shows a current which decreases from an initial "pulsed" level to a DC state. The current decreases with an approximately exponential shape to level off as the measurement time reaches about 800 ns. The lower curve of Fig. 7.1 shows no such reduction of current. Both curves are explained by self-heating. The curves are measured with different drain voltages. The upper curve has higher current and higher voltage than the lower curve and as a result the power dissipated in the channel is nine times larger. This results in a much greater temperature change, and therefore a reduction in current. The lower curve shows that the much lower power does not raise the temperature enough to measurably reduce the current.

This measurement then, is an example of a transistor exhibiting a transient response: from the onset of channel conduction, the measurement records the evolution of current as time progresses. This kind of measurement can be called a transient "waveform" although it is a current waveform, not voltage. Although the applied gate pulse ends at 800 ns, the stimulus condition of this measurement is somewhat like the step function illustrated in Fig. 1.1(c). It is also possible to measure the response to a short pulse, as illustrated in Fig. 1.1(d), and compare it to the DC response, to which the transient waveform extrapolates. In the example of Fig. 7.1, this would mean applying a much shorter pulse, say, 5 ns, thereby excluding the measurement of the time evolution of the current. Both full transient waveforms, and short-pulse measurements can be regarded as measurements of transient properties, but it is helpful to keep the distinction between them in mind, as measurement of full transient response requires techniques different from measurement of short-pulse response.[1] It is noted that ring oscillators would not reveal this kind of change of current due to self-heating: the current in a digital CMOS gate flows only for a few picoseconds and self-heating requires a much longer time.

The important point, however, is that the conventional DC measurement, which has the value shown by the arrow at 800 ns, shows a current which is less, in this case by about 10%, than that which occurs when the transistor is first turned on. As a result, in this technology, DC measurements—the standard practice for characterizing transistors and calibrating models for circuit simulation—do not indicate how much current the transistors will produce in a digital switching situation where the current flows for significantly less than 1 ns. Without one or the other transient measurements, the true current drive of this SOI FET would not be known. Nor do DC measurements suffice for modeling AC or analog circuits: the current available in an AC or analog circuit, where the transistor is always on, may be somewhere between DC and pulsed, according to the power level. The measurement in Fig. 7.1 is also rather coarse in its time scale: one may wonder how much more current could be obtained if the measurement could "zoom in" on the few nanosecond regime. The purpose of the techniques described in this chapter is to measure the current, that is, the transient current, of the transistor immediately after it starts to flow, or in the shortest possible time after its input has changed, to see if it differs from the value obtained from conventional DC measurements. Of course, the time scale for changes might be quite different from the self-heating example of Fig. 7.1, and, as will be seen later, the difference between transient and DC current can be much larger than that which is caused by self-heating.

The ideal instrument for measuring transient response can produce sharp edges or pulses to apply to the transistor and can instantaneously measure the corresponding output current. No such instrument exists.

DC, or static, I–V curves are made with a source-measurement unit using a "staircase" sweep as illustrated in Fig. 7.2 using the example of forcing a voltage and measuring a current. The voltage increases step by step, and at each step, a measurement of current is made. Time is allowed for the voltage to settle before the current is measured, and before the next step is taken, to ensure the measurement is taken in a truly static state. The minimum time for a current measurement is typically about 100 μs, and the voltage step time, 0.5 ms, so a typical measurement sweep will take at least 50–100 ms, and often substantially more, when averaging or a longer integration time is required to reduce noise. Even a *single* point measurement, one step in Fig. 7.2, takes longer than the characteristic self-heating time indicated by the example of Fig. 7.1 and results in a single DC value. Similarly, for

---

[1] As some RF circuits are intended for pulsed operation, such as in radar or communications, some equipment vendors have developed network analyzers which can measure S-parameters during short pulses, e.g., 2 μs to 50 μs, to avoid self-heating. While useful for some applications, these instruments do not lead to measurements of the nature required for rapid transistor time-transient response and are not included in the book.

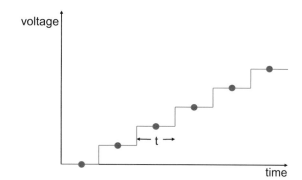

Fig. 7.2 Illustration of the "staircase" method of measuring *I–V* curves employed by conventional SMUs

other phenomena which change the electrical characteristics as a function of time after the voltage is applied, it can be expected that a DC *I–V* sweep takes much longer than the time constant which changes the characteristics, so the transient current is not measurable this way. In some cases, the transistor characteristics might change more slowly than this example, but quickly enough that they change during the full *I–V* sweep. This might cause hysteresis, which was illustrated in Fig. 6.10. DC measurements take a long time mainly because of the current measurement. A source-measurement unit uses some form of op-amp (operational amplifier) circuit to measure current without inserting a load, and the op-amp with feedback, is by nature, a slow circuit. Furthermore, as the current level is reduced, an even longer measurement time is required because of the feedback bandwidth. Auto-ranging is often used to obtain large dynamic range, such as in a subthreshold current measurement, and this switches in different op-amp feedback elements, making the measurement even slower. The problem cannot be solved by forcing current and measuring voltage, as forcing current also requires the use of an op-amp.

7.2   Voltage Measurement Methods

Voltage measurement, on the other hand, can be very fast. The digital oscilloscope, for example, has analog-to-digital converters (ADCs) which, for the most advanced real-time models, have high analog bandwidth (>10 GHz) and digitize at high sampling rates, such as 20 GSa/s, and sampling oscilloscopes have even higher bandwidth. The application of a pulse to a transistor's gate and the observation of its output voltage in a high-bandwidth oscilloscope was discussed in Chap. 6 and illustrated with Fig. 6.3 as a means of obtaining transistor delay. Use of such a voltage in-to-voltage out measurement might seem attractive, but does not lead to a direct measurement of current. It might seem, however, that if the desired current measurement can be converted to a simple voltage measurement, it should be possible to achieve a vast increase in measurement speed compared to routine SMU current measurement.

Two ways of doing this by inserting a sensing element in the current path are suggested in Fig. 7.3. The first method, shown in Fig. 7.3 (a) is to insert a resistor in the current path, such as between the drain terminal of the transistor and the power supply, so the current is determined by the voltage drop, using $V = IR$. The ammeter function of simple digital multimeters (DMMs) works this way. A resistor is switched into the current path, inside the meter, and the voltage across the resistor is measured by the voltmeter. This certainly is a way to measure current, but it has several limitations. The first is what is known as the voltage burden; as a result of the IR drop, the voltage on the transistor is less than the applied voltage and depends on the current and the resistor used. Different resistors are needed for different current ranges so the voltage burden depends on current. Source-measurement instruments are preferred because they do not operate this way, and there is no voltage drop due to a resistor. More important for measuring transients, this technique is not suitable for short transient measurements. A fast measurement can be made with an oscilloscope, but that requires wires to connect to both sides of the resistor. The connection to the oscilloscope must be made with coaxial cable in order to achieve high bandwidth, but in this situation, neither side of the transistor can be grounded, so the cable cannot function properly. A differential input oscilloscope module is needed. Furthermore, the oscilloscope input must be high impedance so the current flows entirely through the load resistor. This leads to high-frequency signals reflecting back and forth on the cable, resulting in severe waveform distortions. So this simple idea is not useful for times of less than about 1 μs. However, a modification of this concept has been proven to be very valuable and will be discussed in great detail below. The sensitivity of this method is equal to the resistor's value. That is, the differential change of voltage across the resistor resulting from the changing current flow, which is measured in the oscilloscope, is described by

$$S = \frac{dV}{dI} = R \tag{7.1}$$

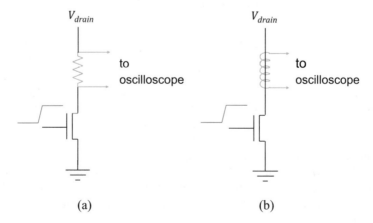

(a)   (b)

Fig. 7.3 Ways to convert drain current to voltage. (**a**) In-line load resistor and (**b**) current transformer

The current resolution is $\delta I = \delta V / S$. For example, if the resistor is 100 Ω and the oscilloscope has a resolution, δV, of 1 mV, then the current resolution is 10 µA.

A second method, as suggested by the illustration in Fig. 7.3(b), is to sense the transient current with a transformer. The transistor current passes through a wire in the center of a transformer sensor. When the current through the wire changes, the change of magnetic flux causes the transformer to generate a larger current which drives a 50 Ω input resistance of an oscilloscope to develop a voltage. Clip-on current-to-voltage transformers, which are conceptually similar to this drawing, are commonly used for measuring 50–60 Hz power currents, though they result in a DC voltage, but there are some, which are commercially available, that are suitable for the small currents and high frequencies of microelectronic circuits [3]. These transformers, by design, are required to drive the 50 Ω input of an oscilloscope, so the bandwidth limits of the inserted resistor technique, discussed above, are not a problem here. Nor is grounding, since the transformer is completely de-coupled from the transistor terminals. The bandwidth is limited, though, by design factors. In one model, the upper bandwidth is 200 MHz, and in another, it is 1 GHz. Since a changing magnetic flux is required for a transformer, this is strictly an AC sensor. It does not respond to DC current, and consequently also has a lower bandwidth cutoff frequency. On the other hand, it does not cause a voltage drop at the device terminal. The sensitivity, S, of commercial high-bandwidth transformers is from 5 mV/mA to 1 mV/mA, substantially lower than the 100 Ω resistor sensitivity cited above, so fairly large currents are required to use such a transformer. The SOI transistor self-heating transient waveforms shown in Fig. 7.1 were obtained using such a current sensor. It is noted that both of the methods of Fig. 7.3 require that the sensor be inserted into a coaxial cable which connects the drain voltage power supply to the microwave probe used for the measurements.

While the current transformer can be useful, as demonstrated by Fig. 7.1, and even the resistor method can be helpful, alternative methods are needed to measure transient currents at very short time scales.

Referring to Fig. 7.2, there are two ways to replace this slow step-and-measure method. The first is to vastly improve the basic speed of measurement, which requires replacing this quasi-static step measurement with very high-speed current measurement circuits, and the other is to abandon this method entirely, and work with short pulses to create and measure only short current pulses. Both methods are discussed in the remainder of this chapter.

7.3   Fast Source and Measurement Units

There are at least two commercial products which address this need for stimulus and measurement at very short times in a fast source and measurement unit (FSMU) [4, 5]. Both have replaced the staircase method with high-bandwidth source and measurement circuits which operate continuously without waiting after a step for the measurement to complete, so they are "on-the-fly" measurements. Both use remote force and sense module to avoid time delays and degradation and reflections from cables, so the measurement circuits are located in small boxes which can be placed very near to the transistors, and the digitized results are transmitted back to the data collection equipment over low-bandwidth cables. They can generate voltage steps or ramps over fairly short intervals, as short as 20 ns, and can be programmed to create arbitrary voltage waveforms, which might be especially useful for novel devices. The current measurement speed, is, however, more limited, and depends on the size of the current. Auto-ranging of the current measurement, which is commonly done automatically in conventional DC SMUs, is not possible, and the selected fixed current range determines the measurement speed. For a range of 10 mA, the time required for current measurement is 100–180 ns, which might be a very long time for some transient processes. For the smaller range of 10 μA, the measurement time increases to 6–10 μs, which may be extremely long compared to some phenomena. Still, these measurement times are far smaller than the typical especially staircase step time of 0.5 ms. More than one fast stimulus and response unit can be operated simultaneously, making it possible to apply fast signals to both gate and drain on an FET. Microwave probes may not be required for measurements on this time scale with the transducers very close, but their use is still advisable to avoid probe impedances causing false artifacts in the measured data.

While these commercial FSMUs are limited in current measurement time resolution, they can be very useful for devices which change characteristics on a medium–short time scale. An example of the use of the rapid measurement SMU to obtain *I–V* curves on a short time scale is shown in Fig. 7.4 [6]. This example shows the drain current of a III–V nFET fabricated with a high-κ metal gate. In the technology studied, the FETs are made with thin InGaAs (indium-gallium-arsenide) channels to understand the effect of thickness of the channel material on electron mobility but the measurements made with a fast SMU showed that conventional measurements

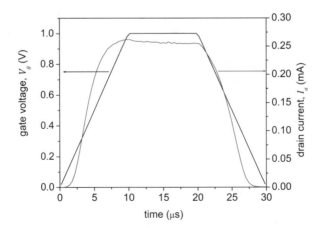

Fig. 7.4 Measured trapezoidal gate voltage applied (black line), and resulting drain current (red line) obtained with a fast SMU, applied to a GaAs nFET

severely underestimate the mobility due to fast charge trapping at the channel-oxide interface. Rather than a step, a trapezoidal voltage, shown in black, is applied to the gate, with a 10 μs rising edge, a 10 μs hold level, and a 10 μs falling edge. The entire measurement shown here take place in less time than a single point measurement of a DC SMU.

These novel transistors are prone to charge trapping at the gate dielectric interface with the channel. Traps, caused by dangling bonds, can charge and discharge at a variety of rates, usually just called fast and slow, and cause changes of threshold voltage. When they are physically near the channel, they can also cause mobility degradation due to scattering from the trapped charge. Evidence of filling of traps is seen in the drain current measurement of Fig. 7.4. The current ramps up in response to the ramped gate voltage, and then can be seen to decrease slightly during the hold time where the maximum gate voltage is constant for 10 μs, and then follows the gate voltage as it decreases. Clearly, there is degradation of current occurring on the time scale of these measurements. This is attributed to fast traps filling during the time that the gate voltage is held constant. From these data it cannot be determined if trapping also occurs at the lower gate voltages as it ramps up from low to high.

These time sweep measurements can be transformed into I–V data by plotting the current vs voltage at every measurement point. The resulting curves are shown in Fig. 7.5(a). There are two typical I_d–V_g curves showing hysteresis in the up vs down measurements and separation of the curves at the highest gate voltage, reflecting the droop in current seen in the time sweep measurement of Fig. 7.4. It can also be seen that the slope of the curves is different for the rising voltage and falling voltage. So, from this single measurement, it can be seen that during operation, the threshold voltage increases, shifting the curve, and the mobility degrades, causing reduced transconductance. The derived transconductance of the two sweeps is shown in Fig. 7.5(b). Such changes of output might not be apparent in a strictly DC measurement, or the degree of the hysteresis observed might be different, as it can depend on the time for which the voltage is applied. In the experiment cited here, it was

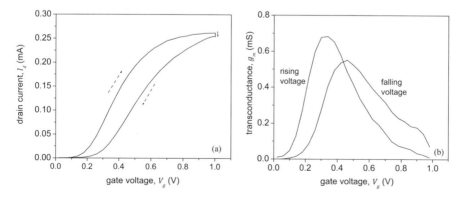

Fig. 7.5 Conversion of the measured drain current of Fig. 7.4 into I–V measurements. (**a**) Transfer characteristic (I_d–V_g curve) and (**b**) transconductance vs gate voltage

reported that a longer hold time led to increased threshold shift but no further reduction of transconductance. The ability to program the voltage waveform of the fast source and measurement units makes it possible to study these time-dependences.

To further study the importance of using short time measurements, comparison with the DC result is always necessary. An example of such a comparison comes from the same InGaAs FET technology in which the intent was to measure the true, that is trap-free, mobility [7]. The measurement is shown in Fig. 7.6. It uses the same measurement conditions as Fig. 7.4, a 10 μs trapezoid, labeled as pulsed measurement, followed by a conventional DC I_d–V_g sweep, followed by a repeat of the trapezoidal measurement. The I–V curve is obtained as in the example of Figs. 7.4 and 7.5, but using only the rising edge of the trapezoid. For comparison, the DC data used a measurement step time of 30 ms. It is seen that the short-time "pulsed" measurements show a much higher output current than the DC. This illustrates how drastically long-time operation of these devices, which allows the fast traps to charge during measurement, can reduce the apparent output current. Higher current is obtained since the traps do not charge as much during the pulsed measurements. The DC curve is affected by both threshold voltage and transconductance. The dashed lines in Fig. 7.6 illustrate how the apparently flattened DC curve is an artifact caused by an increase of threshold voltage during the slow steps of the measurement. As DC steps are applied, the threshold voltage increases, which shifts the intrinsic IV curve to the right. The curve continues to shift gradually to the right as the DC steps continue, and the current measured at each step occurs at a different threshold voltage. It also implies that trapping occurs at the lower gate voltages, but perhaps at a lesser rate.

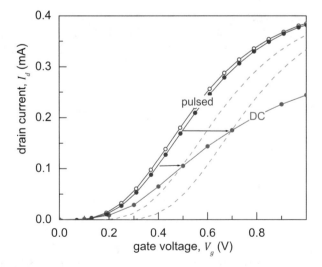

Fig. 7.6 Measured fast ("pulsed") transfer characteristic of a GaAs nFET compared with the same characteristic measured at DC [7]. The grey curves show how the change of threshold voltage during measurement causes an apparent reduction of the slope in the DC measurements

Only measurements on a short time scale can reveal such differences. Furthermore, the almost identical pulsed curves also show that following the trapping during the DC curve, de-trapping occurs quite quickly, and the original state is recovered, when the voltage is removed. By applying two successive trapezoidal pulses and varying the delay between them, it was experimentally determined that the recovery time is on the order of 100 μs to 1 ms. It is further possible to change the ramp rate to try to determine the maximum time required for charge trapping. Again, it is only with the arbitrary waveform generation of fast source and measurement units that such experiments can be tried. However, the ramp rates and measurement times of the fast SMU are on the microsecond time scale. While this scale might be suitable for charge trapping effects, in general it is natural to ask how the novel transistor responds on the nanosecond time scale. For that regime, different measurement methods are required.

7.4   Pulsed *I–V* Method

A short-pulse technique is required to make measurements at much shorter times. In this technique, a very short pulse is applied to the input of the transistor, which is usually the gate, and the output current is measured as a voltage on a high-bandwidth oscilloscope [1]. There is no attempt to measure the time evolution of the transient; rather, the purpose of these measurements is to measure *I–V* curves for short times, to compare them with DC curves. The technique is often referred to as the "pulsed *I–V* method." A series of pulses is applied as illustrated in Fig. 7.7. The pulse can be as short as a few nanoseconds using a suitable pulse generator. A series of pulses of the same voltage is applied, and the oscilloscope measurements made, as will be explained below, and then the voltage of the pulse is changed and another series is applied, thereby generating a sequence of pulses of different voltages, while maintain the same pulse width. The pulses are repeated as long as needed in order to allow averaging of the output to reduce measurement noise in the oscilloscope. The off-time between pulses, where the pulse is at the low level, which is the repetition rate or period, is relatively long, such as 10 μs or more, not closely spaced as drawn in Fig. 7.7.

A diagram of the complete measurement system is shown in Fig. 7.8. As usual, all of the signal connections are made with high bandwidth, 50 Ω cables and components. The pulse generator drives the transistor input, which is the gate, in this schematic, with the desired pulse height and width. No bias tee is needed, but the pulse must swing from a low level which is below the threshold voltage, which turns the transistor off, to the desired high level. A 10× in-line coaxial pickoff probe (tee)

Fig. 7.7 Illustration of the concept of the pulsed-IV method of measuring transient IV curves. The time axis is not drawn to scale

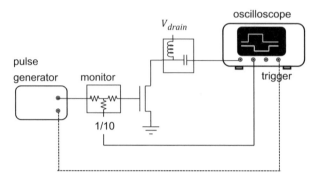

Fig. 7.8 Apparatus used for pulsed I-V measurements. The connections use coaxial cables

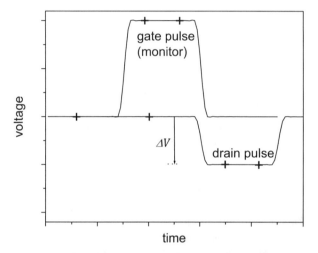

Fig. 7.9 Voltage traces which are seen on the oscilloscope when using the pulsed-IV method

is inserted in the pulse line (and it reduces the amplitude of the gate pulse). This provides a signal which can be displayed on the oscilloscope to monitor the quality (height and width) of the pulse without the need to disconnect and reconnect cables. A separate signal line from the pulse generator is used to trigger the oscilloscope. A bias tee is used to apply a drain voltage. As noted in Chap. 2, passing a pulse through the capacitor can result in a DC shift of the low level of the pulse if the duty cycle of the pulse is sufficiently high, a level which would change the desired drain voltage. Keeping the time between pulses relatively long, so the duty cycle is less than 0.1%, avoids this problem.

The signals of an nFET which are seen on the oscilloscope are illustrated in Fig.7.9. The top trace is the 1/10 monitor of the positive gate pulse, and the lower trace is the drain output. When the gate pulse causes drain current to flow, the current flows from the ground of the oscilloscope to the transistor, through the capacitor, as the short pulse is blocked by the inductor in the bias tee. Since the current flows from the oscilloscope to the drain, and it flows through the 50 Ω input resistance, it shows a negative voltage. (In measurements using a spectrum analyzer, in Chap. 3, in contrast, the *power* dissipated in the 50 Ω input is measured, but has no sign.) But this is more than just an observation of a current across a resistor: it can be used to construct an *I–V* measurement point. This voltage is the measure of the current, similar to the proposed sensing method shown in Fig. 7.3(a), where now the resistor is transferred in space to the oscilloscope, causing a time delay, as seen. It also causes a voltage drop at the device terminal during the time of the pulse.

The AC equivalent circuit of this measurement is shown in Fig. 7.10(b). When the pulsed drain current causes a voltage drop ΔV, as defined in Fig. 7.10(a), the voltage at the drain terminal drops from V_{DD} to $V_{ds} = V_{DD} - \Delta V$. The voltage observed on the oscilloscope therefore determines both the current in the transistor and its corresponding voltage below V_{DD}, i.e., a single measurement results in an *I–V* point:

$$I_d = -\frac{\Delta V}{50} \tag{7.2}$$

$$V_{ds} = V_{DD} - |\Delta V|$$

This analysis is an example of the "load line method" used to solve the problem of a non-linear device driving a resistive load. Shown in Fig. 7.11, the output of the transistor switches from a point on the *x*-axis to intersect the output curve along the load line which has a slope of 1/*R*. For this reason, this pulsed *I–V* measurement is sometimes referred to as the "load line method." It is assumed in this analysis that the drain current is zero before the gate is pulsed, which is why the low level of the pulse generator must be below threshold voltage. Transistors which do not turn off completely, or, more precisely, which do not have a large on/off ratio, cannot be measured with this technique. Repeating this measurement with various values of V_{DD} moves the starting point of the load line along the *x*-axis and generates an I_d–V_d

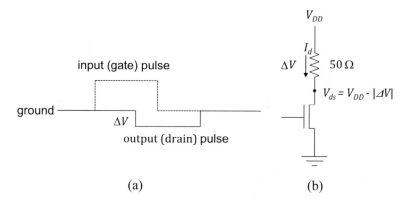

(a)   (b)

Fig. 7.10 AC equivalent circuit of the pulsed I-V method. (**a**) The observed pulse due to the drain current pulse and (**b**) the equivalent circuit from which the current and voltage are determined

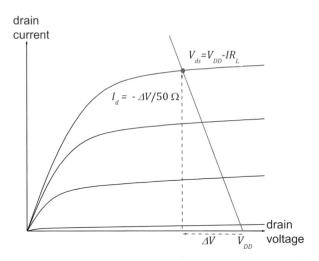

Fig. 7.11 Illustration of the load line concept and how it is used to obtain an I_d–V_d point

curve. Repeating the measurements with different pulse heights generates families of I_d–V_d curves with different gate voltages.

In practice, the measurements can be made by making the oscilloscope measure just a few voltages before and during the drain pulses, as suggested by the tick marks in Fig. 7.9. The gate pulse monitor only has to be measured once per pulse setting. This greatly speeds up the measurement time, as the entire waveform is not acquired. This sample oscilloscope trace is drawn for the case of SOI self-heating where the current does not change substantially during the few nanoseconds of the pulse, but this is not expected to be true for all transistors, and examination of the waveform on the oscilloscope is important, as will be discussed below.

An example of the resulting measurement, shown again for the example of self-heating in SOI FETs, is illustrated in Fig. 7.12. The pulse width is chosen to eliminate or severely minimize self-heating, which can be checked with the waveforms as in Fig. 7.9. The time between pulses is large enough (10 μs in [1]) to allow the transistor to cool off completely between pulses, so there is no accumulated temperature rise due to very small self-heating during the on-pulse. The dots represent the I_d–V_d measurements made by this method, and the solid lines are normal static, DC, curves. The latter can be obtained in the same experimental apparatus by disabling the pulse. In this case, the pulse generator output has the voltage that would be the low level, or base, of a pulse. The drain voltage can be swept, and the current measured, through the bias tee since a DC current is not blocked by the inductor in the tee. The low level of the pulse generator can be varied to obtain curves at various gate voltages. For very careful measurements, the low level may have to be calibrated, as voltage offsets between different equipment often occur. The two superimposed curves of Fig. 7.12 show that, in the illustrated case, the pulsed measurements show considerably more current than do the DC. As the net power in the transistor is higher in the upper right hand corner due to higher voltage as well as higher current, the difference between the two, caused by self-heating, increases,

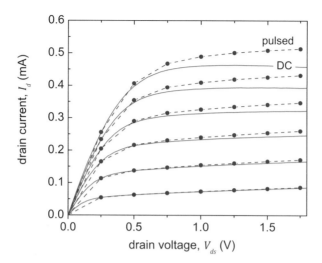

Fig. 7.12 Illustration of the difference between pulsed I-V and DC measurements of output characteristic of an nFET with self-heating

and the slope of the DC curves actually becomes slightly negative as the temperature increases, a sign of higher temperature.

An assumption in these measurements is that the response of the transistor to the repeated short pulses with long waits in between is the same as that of a single isolated pulse, ie, that there is no accumulated change. In general, though, this assumption must be tested. An easy way to do that is to change the time between pulses by a large factor, increasing it to perhaps 1 ms. In the case of self-heating of fully depleted SOI FETs, no difference can be seen between long and very long times between pulses.

While all SOI FETs show self-heating, those with partially depleted bodies also show a floating body and "history effect" due to movement, thermal generation and impact ionization, and recombination of charges in the body of the device which is insulated from the bulk wafer. Unlike bulk FETs, charges can get trapped in the body, and thereby change the threshold voltage. Floating bodies can cause kinks in the DC output characteristics when enough charge is generated to markedly change the threshold voltage. In pulsed measurements, the body charge depends on the interplay of various generation and recombination mechanisms with vastly different time constants [1], so that there is a transistor history-dependence. This has been seen to produce different *I–V* curves for different pulse repetition rates. Unless modifications are made to observe time evolution of the current, it can be safely surmised that a steady-state body charge has been reached by the time of measurement: with GPIB commands taking tens of milliseconds, thousands of pulses are applied to the transistor before the measurements even start.

By using a real-time oscilloscope, however, it is possible to examine if the response to each pulse is the same [8]. The same measurement setup of Fig. 7.8 is used, but the oscilloscope is of the real-time type, and instead of averaging each trace to get the best voltage measurement, each trigger is used to save separate traces. The pulse generator can be programmed to put out a finite number of pulses, and the oscilloscope set to record the same number of traces, without averaging (at the cost of greater noise). A possible result is illustrated in Fig. 7.13. The first pulse applied causes the largest voltage change, that is, the largest current, but at the same time, the charge in the body changes and, in this example, increases the threshold voltage. Subsequent pulses produce less current, and eventually after some number of successive pulses, a steady-state is reached in which the off-state restores the transistor to a certain point so that it always behaves the same. A different time between pulses might result in a different time evolution, and if it is long enough so that the transistor recovers completely, all of the traces will be the same [1]. It is possible, though tedious, to construct a family of successive I–V curves which show the output as a function of time, by using points from the individual traces [8]. FETs with a history effect can be used in circuits, but measurement of the effect of history on their drive capability is required to determine the design margins.

The concept of a pulsed I–V curve is built on the premise that the current doesn't change, as illustrated in Fig.7.9, in the few nanoseconds of the gate pulse. However, there is no guarantee that that will be the situation with any particular transistor. If significant changes occur on the nanosecond time scale immediately after the transistor turns on, then the drain current pulse seen on the oscilloscope will not be flat. This has been seen in floating body SOI nFETs, as illustrated in Fig. 7.14 [1]. When operated well above the threshold voltage, the predominant operating regime, these transistors produce a flat output pulse. But when the gate voltage is only slightly above threshold voltage, the output is not flat when the off-state is sufficiently long. This can be understood as due to two mechanisms which affect the body charge, and hence, the threshold voltage, in an SOI FET, and which occur on vastly different

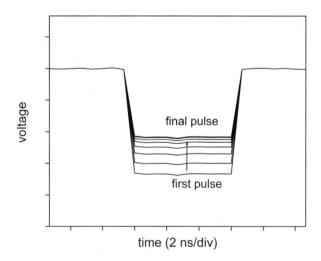

Fig. 7.13 Illustration of the evolution of the drain current pulse in a floating body SOI nFET when the first pulse is applied after a very long off-time [8]

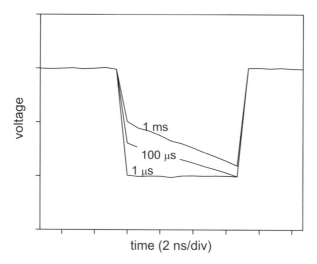

time (2 ns/div)

Fig. 7.14 Illustration of the possible drain current pulse which doesn't saturate in the few nano-seconds of the gate pulse. This occurs for certain bias and timing conditions when the transistor is a floating body SOI nFET [1]

time scales: impact ionization at the drain which almost instantaneously produces charge in the body, and recombination at the source, which slowly decreases the body charge. From data like these it is possible to gain a quantitative understanding of such mechanisms, and the visualization of these currents in an oscilloscope can be very helpful in understanding the dynamic behavior of the transistor.

These measurements are easily made with the linear S-G-S or G-S-G-S-G pad configuration which was proposed in Fig. 3.18 (transistor). Each tested transistor should be isolated from the others; that is, layouts with a single gate, drain or source pad connected to multiple transistors should not be used, as the extra connected transistors may have capacitance large enough to slow down the pulse edges. Because of imperfections there are likely to be some reflections of signals, which limits the length of pulses which can be accurately measured. The longest gate pulse which can be applied before reflections interfere is twice the electrical length of the input signal path, likely to be about less than 10 ns. However, for this measurement, it is possible to use probes with a 50 Ω termination resistor on the probe tips, which can help prevent multiple reflections from connections. The pulse generator level should be programmed to account for the presence or absence of such a termination resistor, as explained in Chap. 2.

The size of the transistor being tested is constrained by upper and lower limits on the current being measured. As described earlier, its sensitivity, S, is equal to 50 Ω, so the current resolution is $\delta I = \delta V/50\Omega$, where δV is the resolution of the oscilloscope, typically slightly less than 1 mV. The transistor must have current substantially larger than this minimum. On the other hand, it cannot be too large, else the load line will exclude a significant portion of the *I–V* curve, as nothing to the right of the 50 Ω load line can be measured without applying a large voltage which might exceed the safe operating limit. Thus, the maximum current of the device is $\Delta V/50\ \Omega$, where ΔV is the maximum acceptable voltage excursion above nominal value for the transistor.

Calibration of the gate and drain signal is required for good measurement. The gate signal suffers a loss at the 10× pickoff, and both gate and drain signal can suffer some attenuation in the high-speed cables and connectors. As can be seen from the load line method, an error in the drain signal voltage measurement causes errors of both current and voltage. Calibration can be performed by putting the same signal directly into the monitor channel and drain channel and measuring their resulting amplitudes. An attenuation factor can be applied to the oscilloscope to correct for any difference, or it can be applied, afterwards, to the pulse measured without the attenuation factor.

Two commercial equipment makers have developed pulsed I–V capability using this overall concept, integrating it into as a module for their device characterization equipment and software. The first [9] uses a minimum pulse of 40 ns, with the pulse having a rise time of 13 ns. As noted, the original method used pulses as short as few ns, with rise times of about 150 ps. The second [10] has capability of a shorter pulse, at 10 ns and sharper rise and fall times, 2 ns. It is claimed to have a current resolution of 1 μA. It also has a compensation method by which the actual drain voltage is adjusted to compensate for the voltage drop due to the 50 Ω load. Unlike the fast SMUs available from the same companies, these pulsed I–V modules can apply and measure just one channel, which is typically the gate.

The pulsed I–V method has been extended to pulses shorter than 1 ns by [11]. Some changes were made to the measuring apparatus, as shown in Fig. 7.15. The 1/10 pickoff monitor was replaced by a power divider, which splits the signal equally between the two branches. The gate probe tip of the microwave probes was terminated with 50 Ω, which was mentioned previously as a possibility. This makes the signal at the gate exactly the same amplitude as the monitor signal at the oscilloscope and prevents reflections of pulses from the gate. The bias tee is replaced by a pickoff tee, which is not described but is presumably custom-designed with resistors which make for impedance matching with the SMU which provides the drain voltage. The conversion from measured voltage on the oscilloscope to drain current

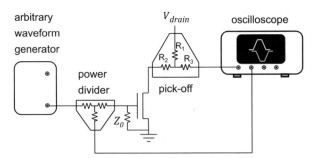

Fig. 7.15 Apparatus used to extend the pulsed I-V method to shorter times [11]

and terminal voltage uses the same load line concept as above, but the load resistance is determined from the combination of the resistors in the pickoff tee as well as the 50 Ω of the oscilloscope.

This measurement system has been demonstrated to measure *I–V* curves with pulses of 200 ns and 500 ps, ten times better than the first pulsed *I–V* experiment, with very short rise times.[2] Applied to study self-heating in SOI 14 nm FinFETs, it showed that the *I–V* curves measured with 200 ns pulses is not much different from DC, whereas those with 500 ps pulses had somewhat higher current. This is consistent with the typical time constant for FinFETs of that dimension, which is on the order of 7–9 ns [12], so that for a 200 ns pulse, the transistor is effectively fully heated.

One advantage which the commercial fast SMUs have, compared to the pulsed *I–V* technique is that the two terminals, typically gate and drain, can both be stimulated with pulsed, trapezoidal, or arbitrary waveforms, though subject to the bandwidth limitations in the Fast SMU section above. The pulsed *I–V* method only allows one terminal, typically the gate, to be pulsed, while the other is held at a DC level. It is by no means obvious that transistor properties are affected by only one terminal experiencing a transient input so it can be desirable to have control over both. For example, it is known that conventional CMOS FETs degrade due to application of voltage due to bias-temperature instability (BTI), even when current doesn't flow through the channel. It might be desired to suppress that degradation by only applying a drain voltage for the short time that the gate is pulsed. The pulsed *I–V* technique prevents this because the inductor in the bias tee intentionally blocks transients from the transistor. Replacement of the bias tee with resistive tee removes this constraint.

A dual-pulse measurement system is shown in Fig. 7.16 [13]. An arbitrary waveform generator is used to produce two independent, but synchronized, waveforms to drive the gate and drain of an FET. The drain voltage pulse is slightly wider than the gate pulse so it reaches and maintains its value well before the gate signal arrives and after the gate pulse has returned to its starting value. Note that the drain terminal of the device is not load terminated, and as, unlike the gate, the drain has a

Fig. 7.16 Apparatus used to apply short pulses to both the gate and drain terminals of a transistor [12]

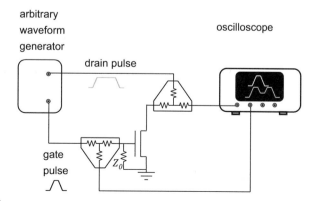

---

[2]The authors state that the edge time is 20 ps, but do not explain how this is achieved.

resistance which varies as the gate voltage controls the channel, the voltage on the drain terminal depends on the voltage on the gate, complicating matters. A complicated but somewhat impractical scheme can be used to extract $I-V$ curves, but the technique suffers from another weakness. The drain is pulsed with a waveform generator, which, as explained in Chap. 2, actually drives a current. Thus, the voltage presented to the input of the drain tee changes as the gate changes the current drawn by the transistor, even during the short time that the drain voltage is applied, making quantitative study difficult. Nonetheless, the technique can be used to study changes, such as those caused by BTI.

7.5 Pulse-Train Method

The pulse-train method is a simple and economical alternate method to obtain pulsed $I-V$ measurements, but can also be extended to be a hybrid between pulsed $I-V$ and full transient waveform measurement [14]. As in the pulsed $I-V$ technique, repetitive pulsed are applied to the gate of the FET. Ideally, the pulses are perfectly rectangular. But instead of using an oscilloscope and the load line method to construct a current, the *average* drain current is measured with a conventional low speed SMU. Thus, no high-speed detection equipment is required although a good pulse generator is needed. This is another example of the general technique mentioned in Chap. 3: stimulate the device with high-frequency or short-time signals, but observe the averaged response using DC measurement equipment.

The measurement concept is shown in Fig. 7.17. All of the connections use coaxial cables. Using a microwave probe to apply signals with good fidelity, a continuous train of pulses is applied and the average current is measured with an SMU,

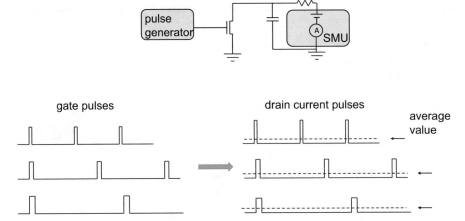

Fig. 7.17 Illustration of the concept and apparatus for the pulse-train measurement technique [13]. <© [2019] IEEE. Reprinted, with permission, from Ref [13]>

which might have an external filter attached. The average DC current measured this way is the average of the off-state and on-state current, so it equals the DC current times the duty cycle of the current pulses. The equivalent full-time current, to compare to DC, is obtained by dividing the measured average current by the duty cycle. By varying the pulse amplitude and drain voltage, *I–V* curves for any given pulse width can be generated. The same apparatus can be used to measure DC curves for comparison. As in the pulsed *I–V* method, by disabling the pulse, the low level of the pulse generator is applied to the gate as a DC level. By sweeping it up and down and measuring the current, DC curves can be traced.

Although the intent is to measure the response to short pulses, it is easy to see that the full pseudo-transient waveform can be observed by using longer and longer pulses, as suggested in Fig. 7.18. The duty cycle must be maintained at a constant value as the pulses are made longer, requiring that the pulse period be changed. In an ideal transistor, the average current measured this way will be independent of pulse width, and average current divided by duty cycle will be equal to the DC value for any width of pulse. In creating a pseudo-transient waveform this way, the measured current is assigned to the midpoint value of the pulse width. This is exactly correct if the current has a linear dependence on time. A refinement of this assignment is discussed below.

A shunt capacitor is added to the SMU output to provide current when the pulses are very long and the currents are large. Without this capacitor, the SMU may not be able to provide the necessary current, or even be driven into oscillation, as it tries to switch quickly and periodically from a long zero current state a high current state. The capacitor provides the extra current if the SMU can't respond quickly enough. A resistor is added to reduce the noise which this capacitor can introduce. This effectively adds a filter to the SMU output. Different SMUs have different *internal* filters for their outputs, so the added capacitor has to be determined by experiment. The indication that a capacitor is needed is a radical and abrupt drop in measured

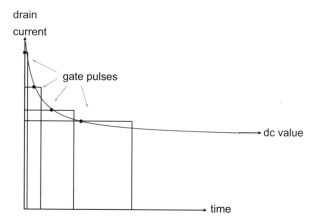

Fig. 7.18 Illustration of the use of pulses of differing widths in the pulse-train method

current when long pulses are used. A capacitor may not be needed at all for small transistors which don't draw much current.

Unlike the load line method, the output from the device is not observed on an oscilloscope, and the input pulse is not observed via the monitor pulse. This makes it imperative to determine the quality of the input pulse using auxiliary measurements. The basic measurement system can be tested using a simple short, low resistance metal strip on a substrate, where the strip is used to connect the gate probe to the drain probe. Setting the drain voltage to zero grounds the voltage pulse, creating a current in the SMU. The amount of this current can be predicted from the total load resistance encountered by the pulse, which is the sum of the metal resistance and the resistance of the external SMU filter, if installed. The current is not equal to the voltage of the pulse divided by the resistance, as transmission line reflections affect the voltage at the termination. Unless the resistance is equal to the transmission line impedance, Z_0, the voltage will be changed. From this consideration, it can be shown that

$$V_L = V_i\left(1+\Gamma\right) = V_i\left(\frac{2R_L}{R_L + Z_0}\right) \tag{7.3}$$

$$I_L = \frac{2V_i}{R_L + Z_0}$$

where V_L is the voltage at the load and I_L is the corresponding current, and V_i is the input voltage.

An example of measuring the current through a gate-to-drain metal connection is shown in Fig. 7.19 for two different pulse generators. Continuous trains of pulses from as short as 2 ns to as long as 100 µs are applied to the SMU through the on-chip resistive connection with the microwave probes, and the average current measured. Both pulse generators show that, on average, the measured DC current is the same for all pulse widths, as expected if duty cycle is kept constant. Pulse generator 1 shows some departures from the average for some pulse widths. Observing the pulses with an oscilloscope, it was determined that their widths, hence, their duty cycles, varied slightly from the programmed value. In this case, a correction table can be constructed and used to correct the data, resulting in the data which are superimposed on the original in Fig. 7.19. As the connection from the pulse generator to the SMU goes through microwave probes, these measurements also show that the probes do not impose any bandwidth limits or distort the pulses.

The effect of different duty cycles is shown in Fig. 7.20 using pulse generator 2, and using the same connection described for Fig. 7.19. The data show that the measured current, as above, is constant over a very wide range of pulse widths, and that the current is proportional to the duty cycle of the pulse. There is a very slight difference in the ground levels between the pulse generator and the SMU which causes current flow even when there is no pulse, so there is an offset of the currents which accounts for the apparent mis-scaling between 5% and 10%. The inset shows that extrapolation of duty cycle response does not intersect at zero. Such a difference in

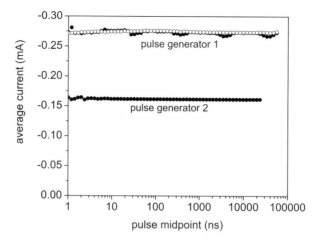

Fig. 7.19 Measured verification of the pulse-train method using an on-wafer resistive connection of gate and drain probes and 5% duty cycle [13]. The current resulting from two different pulse generators are shown. In the case of pulse generator 1, the measured currents are also shown corrected for slight difference between of the measured and programmed duty cycle. <© [2019] IEEE. Reprinted, with permission, from Ref [13]>

grounds is significant in a resistive connection, but is of little consequence, if small compared to gate voltage, when the pulse train is applied to a transistor which has a threshold voltage which must be exceeded before current flows.

A verification of the pulse-train method to measure pulsed device current is shown in Fig. 7.21, where the transistor measured is a conventional SOI nFET. As in the tests described above, pulses are varied from 2 ns to 100 μs, and operated at a 5% duty cycle. The pulse generator used was pulse generator 1, with the correction table applied, and the pulses have rise and fall times of less than 200 ps. The measured current of the FET is approximately constant over the wide range of pulse widths, as expected. The line drawn in Fig. 7.21 indicates the value of 5% of the measured DC current. Within errors, it has the same value as the pulsed measurement, demonstrating exactly what is expected for a transistor which behaves at short times the same way as it does at DC.[3] As in the pulsed I–V method, the transistor can use the S-G-S or G-S-G-S-G layout and probes of Fig. 3.18 (device) and the gate probe can be terminated on the probe tip with a 50 Ω resistor.

The current starts to drop a little, less than 3%, for pulse widths less than 10 ns. No such drop is seen the calibration studies of Figs. 7.19 and 7.20. While this may lead to the conclusion that the device behaves differently for these pulses, it actually illustrates another quality of the pulse which needs to be considered in using this technique. The rise and fall times of the pulses account for this difference. An ideal

---

[3] Unlike the examples shown in the Pulsed I–V section, the current in these SOI FETs is not sensitive enough to change due to self-heating.

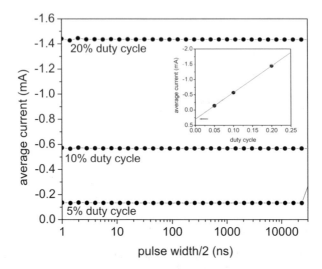

Fig. 7.20 Measured current showing the effect of duty cycle using an on-wafer resistive connection of gate and drain probes, using pulse generator 2 [13]. Inset: current *vs* duty cycle and extrapolation to zero duty cycle

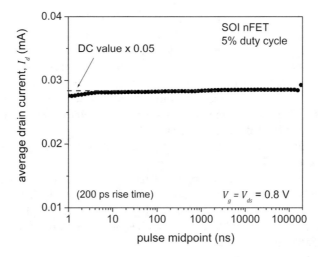

Fig. 7.21 Result of applying the pulse-train method to a conventional SOI nFET. Pulse generator 1 was used with a 5% duty cycle and the correction for duty cycle error was applied [13]. <© [2019] IEEE. Reprinted, with permission, from Ref [13]>

pulse is rectangular, but in reality, all pulses have somewhat trapezoidal shapes, as illustrated in Fig. 7.22

The current through a resistor when a pulse is applied is equal to the area of the pulse. If the width of the pulse is the time between the half-heights of the rising and falling edges, then, for current through a resistor, the pulse is equivalent to a perfect

Fig. 7.22 Illustration of the effect on pulse generator rise and fall times on the effective duty cycle in the pulse-train method

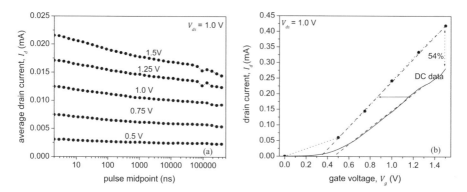

Fig. 7.23 Measured transient response of GaAs nFET using the pulse-train method with 5% duty cycle [13]. (**a**) Drain current *vs* pulse midpoint at various gate voltages and (**b**) I_d–V_g transfer curve extracted from this data for the 4 ns pulse, normalized by duty cycle, compared to the DC-measured curve. <© [2019] IEEE. Reprinted, with permission, from Ref [13]>

rectangle of the same width, i.e., a rectangle of the same area as in Fig. 7.22(a). However, when applied to a transistor with a conduction threshold which is above the low level of the pulse, only that portion of the pulse above the threshold contributes to the current, e.g., Fig. 7.22(b). (However, a serious voltage offset of the pulse above the desired low level means that the gate voltage applied might be wrong, so a calibration check is advisable.) Thus, the *effective* pulse width for the transistor is less than the nominal width. When pulses are long, this difference is very small, but when pulses are short, the effect can become significant, as the SOI FET example of Fig. 7.21 illustrates. This underscores the importance of using the highest quality pulse generator possible and understanding its edge-rate limitations.

An example of using the pulse-train method to measure short-time characteristics of non-conventional FETs is shown in Fig. 7.23 [14]. The transistor tested was a III–V (InGaAs) long channel nFET. The measurements were made as in the SOI nFET (Fig. 7.21), but for this transistor, the height of the pulse, that is, the gate voltage, was set at several different values, resulting in separate waveform curves, as labeled in Fig. 7.23(a). It is seen that for the larger gate voltages, there is a marked difference in drain current between the short pulses and the long ones. It is also noted that for this FET, the longest pulses are not long enough to show the current reaching the asymptotic DC value, and so DC data have to be obtained by disabling the pulse, as described above.

Figure 7.23(b) compares an I_d–V_g curve using DC measurements with a short-pulse curve obtained from the 4 ns data taken from Fig. 7.23(a). At the highest gate voltage, the normalized short-pulse current is more than 50% greater than the DC value. The linear extrapolation of each I_d–V_g curve to zero current also shows that DC conditions increase the threshold voltage by more than 100 mV. These observations are attributed to oxide traps which are prevalent in III-V materials. In DC operation, the influence of traps cannot be resolved if the trapping is fast. Furthermore, the occupation of traps depends on the gate voltage, causing different current-*vs*-time slopes seen in Fig. 7.23(a). The short-pulse technique clearly shows the capability of assessing the *intrinsic* performance by using short time conditions where trapping is avoided or negligible.

A much more extreme example of the difference between short-pulse and DC output is shown with the example of an FET using MoS_2 as the channel material. MoS_2 is a two-dimensional material, and as used for the device channel, it is essentially all surface and has no bulk interior. This makes it, and other two-dimensional materials such as graphene, potentially very sensitive to environmental contamination and to charge trapping at its interfaces. The MoS_2 FET is made by the simple method of depositing a flake on an insulating substrate, using the structure depicted in Fig. 3.1(b), and as a result, it is not encapsulated to protect it from the environment.

The DC I_d–V_g characteristic is shown in Fig. 7.24(a) for three successive measurements of one transistor. Even these three DC curves indicate instability of the output, as there is a difference between them of more than 20%. The pulse-train measurement in Fig. 7.24(b) shows an even more dramatic effect. The short-pulse measurements show a duty cycle corrected current which in one case is greater DC by a factor of almost 10. This shows the importance of short-pulse measurements with novel devices. The performance of this transistor is so greatly underestimated by the DC measurement that it might be considered useless, if it were not for the performance revealed by the transient pulse measurement. Also, the variation

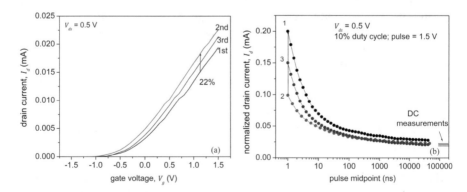

Fig. 7.24 Measurement of transient response of MoS_2 nFET using pulse-train method. (**a**) DC-measured transfer characteristics and (**b**) drain current *vs* pulse midpoint measured with 10% duty cycle pulses. The average drain current has been normalized by the duty cycle for comparison with the DC value

between the different measurements is amplified by short-pulse test, as the minimum and maximum pulsed outputs vary by 100%.

The importance of *very* short measurements is shown by the replotting of the Fig. 7.24(b) data on a linear time scale in Fig. 7.25. Although it is apparent on a logarithmic time plot, the linear plot emphasizes that the dramatic increase of current only occurs when the pulse is less than about 100 ns. Because the current increases sharply only at very short time scales, the integration of the signal over a long measurement window fails to show the dramatic increases. A fast SMU, for example, which integrates, under good circumstances, over a 1 μs window, would see only a 40% increase of current due to the very narrow enhancement instead of a 600% increase revealed by one measurement by the pulse-train method shown in Fig. 7.25.

When such dramatic changes of current for different pulse widths are seen, it is worth considering adjusting the assignment of time to the current. As described in the beginning of this section, the time associated with the current measured is taken to be the midpoint of the pulse. This is correct when the current changes linearly, or approximately linearly, over the time interval. If only the shortest pulse is required, it is a satisfactory approximation, but if the transient waveform is desired, Fig. 7.25 shows a response that is far from linear over long pulse times, and an improved assignment method is needed.

A simple improvement can be made by using the difference between currents measured for two pulse widths. The subtraction is illustrated in Fig. 7.26. Two pulses are shown, w_1 and w_2. The total current measured in w_2, I_2, includes that of w_1, I_1, and is plotted at $w_2/2$. As the current in w_1 is rising rapidly, this clearly overestimates the current. This can be remedied by subtracting I_1 from I_2, and using the midpoint of the $w_2 - w_1$ block, which is $(w_2 + w_1)/2$, as the assigned measurement time. Because the measured currents are means, they are integrals of the current as

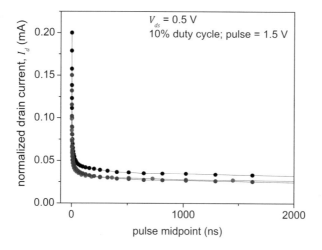

Fig. 7.25 Data of Fig. 7.24(b) plotted on linear time scale

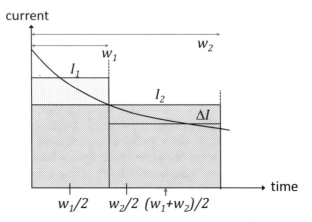

Fig. 7.26 Illustration of the method to correct the time-placement of data in the pulse-train method

a function of time divided by the pulse width. The difference of currents, ΔI, is obtained by subtracting the integrals over the two widths, and then taking the mean by dividing this difference by the difference of the widths. Representing the integrals as rectangles (an approximation, to be sure) this results in:

$$\Delta I \left(w_2 - w_1 \right) \cong w_2 I_2 - w_1 I_1 \tag{7.4}$$

whence

$$\Delta I \cong \frac{w_2 I_2 - w_1 I_1}{w_2 - w_1} \tag{7.5}$$

which is the current assigned to the measurement time, $(w_2 + w_1)/2$.

This reassignment of the data using differences is demonstrated for one curve of the MoS$_2$ FET transient response in Fig. 7.27. The current difference, as defined above, is determined from the raw measurements, and plotted against the revised midpoint, represented as the open circles. The shape of the response is changed somewhat, predominantly in the regime below 100 ns. The subtraction introduces some measurement noise, but does result in a better approximation to the true transient response. The relatively small change suggests that the adjustment is unlikely to make much difference when the short-pulse and DC currents are closer in value than in this example.

The pulse-train method is very easy to implement and has advantages and disadvantages compared to the load line pulsed I–V technique. In many ways, it is the best of all methods for measuring transient response. It is, first of all, very simple. It can be used for short-pulse measurements and for full transient waveforms. There are almost no transistor size limitations, as there are for the pulsed I–V method. Because of the very low current measurement capabilities of SMUs, (especially compared to oscilloscope noise floor) very small transistors can be measured

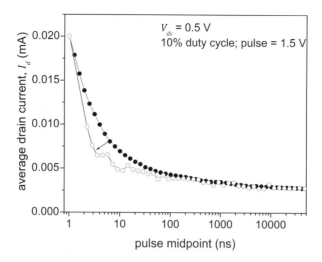

Fig. 7.27 Average pulse-train drain current *vs* pulse midpoint corrected with time-placement method described in text. The black dots are the original measurements and the open red circles are corrected by the difference method

although the duty cycle must be large enough to keep the averaged signal above the noise floor. And since there is no load line to exclude portions of the output characteristic, transistors are not subject to the upper size limit. The time resolution is determined by the pulse generator and not by the measurement itself; improved pulse generation can push the measurement to shorter time scales.

However, the method is explicitly dependent on the application of a continuous train of pulses. The use of continuous pulses implies, as was pointed out before, that the transistor recovers to its original state during the off-time between the pulses. The off-time depends on the pulse width, since duty cycle is held constant when the width is varied, so this recovery condition is not guaranteed, and is a potential problem for very short pulses. In fact, the condition measured is that of a steady-state, which state might depend on pulse width or duty cycle. That dependence can be checked by experimental measurement. The off-time cannot be made too long, however, as that reduces the average current, which might, for a small transistor, lead to measurements near the noise limits of the SMU. Another disadvantage is the fairly long measurement time. Averaging is required at each setting of the pulse and period parameters, and a full-time sweep combined with voltage sweeps quickly adds up to a large number of measurements. Last, the measurement by the SMU integrates all of the output current during the pulse, and since there is no visualization of this pulse, as there would be with an oscilloscope, unexpected transient effects, such as the evolution of current with pulse history seen in Fig. 7.13 and the time development at a short time scale as in Fig. 7.14, cannot be observed, although the latter can be deduced by using many short pulses differing only slightly in width.

7.6   Fast Edge Methods

Seeing how the response of a transistor can depend very strongly on the duration of its input signal, it is natural to desire to have test methods which can extend to even shorter times than shown in the preceding sections of this chapter. This final section describes two techniques which make use of the fast *edges* of an input pulse to extend techniques which have already been demonstrated above.

The first case examines the degradation which can result from charging of fast traps. In the Fast Source and Measurement Units section above, measurements showed, as in Fig. 7.4, that the output current of a transistor can degrade soon after the channel starts to conduct because of charge traps becoming filled. It is well known that traps can have a wide variety of time constants for filling and emptying, and the speed with which they fill determines the effect on output current, according to the rapidity with which the channel turns on. In Fig. 7.4, the gate was ramped from its off- to its on-state in 10 µs, after which the current degraded, implying that traps continued to fill after the current was ramped up in 10 µs. But it is natural to ask if there are faster traps which are filled during the up-ramp of the gate. If so, that would imply that the current available in a picosecond or nanosecond switching event could be higher than what is measured even on a microsecond time scale. Though framed in terms of traps in a InGaAs FET, it is natural to ask for measurements on a shorter time scale for any novel device which exhibits a difference of output between modestly short time scale and DC.

To attack this question requires use of a much faster input signal than a fast SMU can produce or measure. The pulsed *I–V* method addresses this question, but pulses are limited to being at least as long as a few nanoseconds. However, by giving up the fast measurement feature of the fast SMUs and using oscilloscope voltage measurements, converted to current, it is possible to apply a fast-rising gate signal to examine the response very soon after the channel is quickly turned on. Similar to the pulsed *I–V* method, current is observed as a voltage on a 50 Ω oscilloscope. The voltage is measured during the rising edge and following flat portion of a pulse applied to the gate [15], thus constituting a transient current measurement. As in the pulsed *I–V* method, the transistor must provide enough current to develop a large voltage in the 50 Ω oscilloscope input. The apparatus for this measurement is the same as previously shown in Fig. 7.8 or Fig. 7.15. The demonstration experiment was done using an arbitrary waveform generator as in Fig. 7.15. The transistors studied were Ge pMOSFETs fabricated with an Al_2O_3/GeO gate stack. This channel-dielectric system was thought to have interface traps with time constants of 100 ns or smaller, requiring measurement on a short time scale.

The arbitrary waveform generator in the experiment is able to produce edges of programmable rise times, as short as 500 ps, more than two orders of magnitude faster than those available with fast SMUs. The transistors used in the demonstration produce about 1 mA of current, giving a sizable voltage on the oscilloscope. The results are shown qualitatively in the drawing of Fig. 7.28. Gate voltages of rise times of 500 ps and 10 ns are applied, as indicated by the idealized dotted lines. The

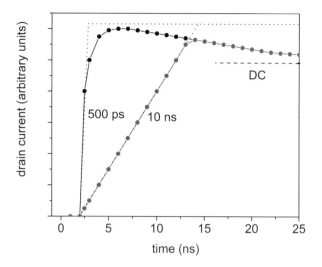

Fig. 7.28 Drain currents measured with fast edge techniques, for edges of two different rise times [14]

actual gate waveform at the transistor input is *not* measured. The output is measured as voltage on the oscilloscope and converted to current. With oscilloscope sampling rates as high as 100 GSa/s (10 ps per point), even a signal as short as 500 ps can be digitized with a large number of points. The measurements show that for the shortest rise time, the current decreases within a few nanoseconds of the channel turning on, and shows a decline which appears to be leading asymptotically to the value which would be measured at DC. The response to the 10 ns edge similarly shows an initially higher current which degrades with time, but it is clear from the time-dependence that a significant degradation, due to traps filling, also occurs during the 10 ns required to raise the gate voltage. This is a strong demonstration of the very fast trapping which occurs in this particular dielectric-channel interface, indicating again, that DC measurements, and even microsecond measurements, are pessimistic predictors of potential digital speed. A complication is noted: the drain voltage decreases as the current increases, and although this reduction does not exclude portions of the I_d–V_d curve, as in the pulsed I–V method, it does suggest that a very large drain current will result in a substantially lower drain voltage at the transistor terminal.

The second case extends the technique of obtaining an I–V curve from fast rise and fall time measurements to short time scales. Fast SMU measurements showed the extraction of I_d–V_d curves from a fast gate ramp in the measurements shown in Fig. 7.4 and 7.5 above. In that example the fast SMU measured the input voltage and the output current as a function of time. Extending this method to fast voltage edges, the current is again measured as a voltage on the oscilloscope, as in the apparatus depicted in Fig. 7.15. The fast gate voltage ramp comes from a pulse generator or arbitrary waveform generator, and the oscilloscope has high sampling rate, so simultaneous gate voltages and drain voltages can be digitized, and the drain voltage converted to current, as before [16]. It was remarked in the section on Voltage Measurement Methods that current cannot easily be obtained from voltage-to-voltage measurements, such as in Fig. 6.3. However, this is precisely what is done when the voltage on an oscilloscope is converted to current, so long as it is recognized that the drain voltage is simultaneously reduced by the load. The result is data like the rising edges in Fig. 7.4, but with a much faster rise time. By plotting the drain current *vs* the gate voltage, rather than *vs* time, the I_d–V_d response is obtained as it was in Fig. 7.5(a). The technique has been demonstrated with 14 nm CMOS finFETs for the purpose of rapidly measuring the change of threshold voltage caused by bias-temperature instability (BTI) stress.

In that work, the gate rise time is 700 ps, more than four orders of magnitude faster than that of Fig. 7.4. While this fast edge technique is a big advance over the fast SMU and pulsed I–V methods, it has raised some potential concerns. First, as has been mentioned repeatedly, measurement in a 50 Ω oscilloscope requires that the drain current be large enough to be greater than the instrument's noise level, so it may not be suitable for some novel transistors with low currents.[4] Second, the gate voltage and drain voltage traces have to be measured with the same time delay on the oscilloscope. In Fig. 7.4, any difference in path delay of a few nanoseconds is insignificant, but it is easy to see that if the drain signal is not synchronized to the gate signal, when the entire I–V curve is measured in less than 1 ns, the timing error results directly into a V_t error. This error is due to the unknown time difference between the gate and drain signals, and is greatest when the rise times are shortest.. Time de-skewing can be performed by the oscilloscope, so a passive calibration device can be used to measure the path delay difference to apply to the oscilloscope. Third, the gate voltage measured by monitoring the divided signal which is sent directly to the oscilloscope may differ slightly, but significantly, from the signal which arrives at the gate terminal, but is not itself measurable. The signal actually applied to the gate goes through a different cable, through the microwave probe, and has to drive the pad and gate capacitance of the transistor, possibly reducing its rise time or changing its shape so that using the monitor voltage to obtain the I_d–V_d curve does not correctly represent the true response. Last, because of the load line effect, the drain voltage is changing throughout the I_d–V_d measurement, so it is not the same as a DC I_d–V_d curve where constant V_d is applied. This is not a problem with the active amplifier measurement of the fast SMU, but is an inevitable result of measuring current with a load resistor. It is not necessarily a problem in looking at only transient effects, but it must be taken into account if the curves are compared to DC curves.

The use of the fast edge from a waveform generator combined with the load line method of measuring current introduces interesting possibilities for measuring transients and I–V curves on a sub-nanosecond time scale. While there have been

---

[4] It must be pointed out though, that some novel devices may have much larger transient currents than what would be apparent from the DC measurement, as in, for example, Fig. 7.25.

successful demonstrations, there remain concerns about the accuracy of the results, and the techniques are probably best suited as indications of rapid processes, and not necessarily as absolute measurements.

References

1. K. A. Jenkins, J. Y. Sun and J. Gautier, "Characteristics of SOI FET's under pulsed conditions," *IEEE Transactions on Electron Devices*, Vol. 44, vol.. 11, pp. 1923-1930, Nov. 1997, DOI: https://doi.org/10.1109/16.641362.
2. L. T. Su, J. E. Chung, D. A. Antoniadis, K. E. Goodson and M. I. Flik, "Measurement and modeling of self-heating in SOI nMOSFET's," *IEEE Transactions on Electron Devices*, vol. 41, no. 1, pp. 69-75, Jan. 1994, DOI: https://doi.org/10.1109/16.259622.
3. CT1, CT2, and CT6, products of Tektronix, Inc, Beaverton, OR, www.Tek.com
4. 4225-PMU, product of Keithley Instruments, Tektronix Inc., Beaverton, OR, www.tek.com.
5. B1530 WGFMU, product of Keysight Technologies Inc., Santa Rosa, CA, www.keysight.com.
6. E. Cartier, personal communication
7. E. Cartier *et al.*, "Electron mobility in thin In0.53Ga0.47As channel," *2017 47th European Solid-State Device Research Conference* (ESSDERC), Leuven, 2017, pp. 292-295, DOI: https://doi.org/10.1109/ESSDERC.2017.8066649.
8. K. A. Jenkins, Y. Taur and J. Y. -. Sun, "Single pulse output of partially depleted SOI FETs," *1996 IEEE International SOI Conference Proceedings,* Sanibel Island, FL, USA, 1996, pp. 72-73, DOI: https://doi.org/10.1109/SOI.1996.552499.
9. Model 4200-PIV, product of Keithley Instruments, discontinued.
10. B1542, Ten Nanosecond Pulsed IV Parametric Test Solution, product of Keysight Technologies Inc., Santa Rosa, CA, www.keysight.com.
11. Y. Qu et al., "Ultra fast (<1 ns) electrical characterization of self-heating effect and its impact on hot carrier injection in 14nm FinFETs," *2017 IEEE International Electron Devices Meeting (IEDM)*, San Francisco, CA, 2017, pp. 39.2.1-39.2.4, DOI: https://doi.org/10.1109/IEDM.2017.8268520.
12. F. Stellari, K. A. Jenkins, A. J. Weger, B. P. Linder and P. Song, "Self-Heating Measurement of 14-nm FinFET SOI Transistors Using 2-D Time-Resolved Emission," *IEEE Transactions on Electron Devices,* vol. 63, no. 5, pp. 2016-2022, May 2016, DOI: https://doi.org/10.1109/TED.2016.2537054.
13. Yiming Qu, Bing Chen, Wei Liu, Jinghui Han, Jiwu Lu, Yi Zhao, "Sub-1 ns characterization methodology for transistor electrical parameter extraction," *Microelectronics Reliability*, vol. 85pp. 93-98, June 2018, DOI: /https://doi.org/10.1016/j.microrel.2018.03.022.
14. K. A. Jenkins, E. A. Cartier and J. Yau, "Pulse-Train Method to Measure Transient Response of Field-Effect Transistors," *IEEE Electron Device Letters*, vol. 40, no. 2, pp. 171-173, Feb. 2019, DOI: https://doi.org/10.1109/LED.2018.2887007.
15. X. Yu et al., "Quantitative Characterization of Fast-Trap Behaviors in Al2O3/GeOx/Ge pMOS-FETs," *IEEE Transactions on Electron Devices*, vol. 65, no. 7, pp. 2729-2735, July 2018, DOI: https://doi.org/10.1109/TED.2018.2836398.
16. X. Yu et al., "A Fast V_{th} Measurement (FVM) Technique for NBTI Behavior Characterization," *IEEE Electron Device Letters*, vol. 39, no. 2, pp. 172-175, Feb. 2018, DOI: https://doi.org/10.1109/LED.2017.2781243.

Appendix A: Measuring Transistors with Controlled Temperature

Although controlling the temperature of the transistor is not a measurement technique in itself, and is therefore outside of the stated realm of this book, it deserves some consideration. Semiconductor devices usually have significant temperature dependence, and novel devices should be examined to determine their temperature behavior. Properties such as mobility and impurity activation are expected to depend on temperature in known semiconductors, but novel effects like transient response may have surprising results at high or low temperatures.

Sometimes it will be sufficient to measure DC properties at various temperatures, including room temperature, and relate them to the results of measurements described in this book measured only at room temperature, but more likely, it will be necessary to repeat these advanced measurements at other temperatures.

The use of microwave probes for most of the techniques presented here can present some difficulty in temperature-dependency studies, particularly for extremely low temperatures. The purpose of this appendix is to point out some of the problems which may occur. While microwave probe have eased the effort of high-frequency measurements enormously, their use in temperature-controlled environments is not so easy.

Modern probe stations can be equipped with temperature control of the wafer chuck. A typical available temperature range covers the military range, from −55 °C to +125 °C, and sometimes extends to 200 °C.

Measuring at temperatures above 0 °C presents only a few problems. First, the probe must be rated for use at the maximum temperature required. Some probes contain organic materials which might melt or soften when hot. Second, the probe pads must be made of a metal which doesn't oxidize rapidly at high temperature. Aluminum and gold are excellent, but other metals, such as copper, will quickly grow insulating oxide layers when heated. Third, it is important to wait for the probe station to stabilize after the chuck is at the target temperature. The heater is in the chuck, but after the chuck is hot, the chassis which holds it also gradually warms up and expands, which may cause a small movement of the chuck, in both the horizontal and vertical directions. If this occurs when the probes are in contact, the probes

© Springer Nature Switzerland AG 2022
K. A. Jenkins, *RF and Time-domain Techniques for Evaluating Novel Semiconductor Transistors*, https://doi.org/10.1007/978-3-030-77775-3

may be damaged, and they may damage the probe pads, and probably lead to erroneous measurements, if contact is lost. The probes themselves also heat up from radiative and conductive heating, and some time should be allowed for their temperature to change.

Measuring at temperatures below 0 °C introduces additional complications. The mechanical problem of contraction is similar to the problem of expansion at high temperature, but many difficulties arise from the fact that ice will form on the wafer from water vapor in the air. The formation of ice on the wafer is unacceptable for many reasons, so measures must be taken to prevent it. The method used for conventional probe stations is to seal the wafer chamber from the atmosphere as much as possible and flow nitrogen with positive pressure into the sealed environment. This flow displaces the ordinary water vapor laden atmosphere.

The sealed environment is established by building a small enclosure, which covers the usually exposed area of the wafer, and provides a hole for the microscope to pass through, slots for probe manipulators, and cables to enter the encloser. These entry points are filled with flexible gaskets to perform some sort of seal around the cables and manipulator. These gaskets, if they are to seal well, tend to resist the motion of the manipulators or cables. Therefore, the motion of the probes is very limited, and the position of the probes, microscope, and wafer should be established before closing the enclosure and cooling the chuck. After testing, the chuck should be warmed up before removing the wafer, to avoid frost and condensation accumulation on the wafer and probe station mechanical parts.

Operation at temperatures lower than −55 °C requires the use of a dedicated cryogenic probe station, or immersion of a packaged sample in the cryogenic liquid or closed-cycle cryostat. (The packaged sample solution is the best method of temperature control, but in addition to requiring fabrication of high bandwidth package, it drastically limits the number of transistors which can be evaluated.) Cryogenic probe stations operate by circulating a cryogenic fluid in the chuck, where the two cryogenic fluids commonly used are liquid nitrogen, which has a boiling point of 77 K, or −196 °C, and liquid helium, with a boiling point of 4.2 K. Some probe stations can operate with either liquid, while some are limited to the warmer liquid nitrogen. Other chuck temperatures are possible if a heater and temperature controller are part of the system, which is fairly standard.

Because the cool down requires cooling the large volume that comprises the wafer chuck which is a large thermal mass, the cool down time depends on the chuck diameter. To cool a 300 mm chuck to liquid helium temperature would require more than 12 h and would consume an extraordinary amount of the liquid. For this reason, most cryogenic probe stations have very small sample stages, much less than the 300 mm standard for conventional CMOS wafers. This is not a problem for experimental transistors built on small substrates.

In order to prevent ice forming at such extremely low temperatures, the cryogenic probe stations use a well-sealed sample chamber, which is evacuated before cooling down. In ordinary probe stations, the sample substrate is held to the chuck by vacuum, but in the vacuum of a cryogenic probe station, some sort of mechanical clips must be used to fasten the sample substrate to prevent movement when contacted by the probes. Some practitioners use, in addition, a layer of a soft thermal conducting material, such as indium, between the substrate and the chuck, to insure a good thermal contact.

To prevent heat from entering the chamber, the probes are manipulated through long flexible bellows which reduce the heat flow and maintain the vacuum. As a result, the probes must be mounted on long arms which may multiply small vibrations and movement outside of the chamber. This can lead to intermittent contact with the sample, and a good thick probe pad metal is advised, to avoid this problem.

Finally, the microscope has to be kept outside of the vacuum chamber, and the sample viewed through a small window. This imposes additional constraints on the microscope. As it cannot be close to the sample, an objective lens with a long working distance is required. Due to that, vibration of the microscope can make viewing the sample difficult, and a very rugged mounting system is needed to minimize vibration. Also because of the distance, adjusting the microscope to view the sample and the probes simultaneously can be difficult. It is best to establish positions of both before the chamber is sealed and evacuated, and before the sample stage is cooled.

In the case of liquid helium operation, it is very likely that the heat flowing from the room environment to the probe body will raise the sample temperature above the chuck temperature by as much as 5 or 10 K. If this is a concern, it is advisable to attach a temperature-measuring sensor on the sample wafer so that the temperature rise can be monitored. It is also possible to purchase special cryogenic RF probes whose bodies are cooled by attaching a metal braid to the sample stage, bringing them to the same temperature.

Index

© Springer Nature Switzerland AG 2022
K. A. Jenkins, *RF and Time-domain Techniques for Evaluating Novel
Semiconductor Transistors*, https://doi.org/10.1007/978-3-030-77775-3

Printed in the United States
by Baker & Taylor Publisher Services